Protect Your Business in the Digital Age

A Non-Technical Guide to Cybersecurity for Small Business Owners

Roger Best

CONTENTS

INTRODUCTION

WHY CYBERSECURITY MATTERS TO EVERY SMALL BUSINESS

P icture this: it's a perfectly ordinary Tuesday morning. The kind of morning where you think, "Yes, today will be productive. Today I will tackle my inbox, drink my coffee while it's still warm, and maybe even—if the universe is feeling especially benevolent—find time for a proper lunch." Everything is ticking along nicely. The coffee isn't quite as hot as you'd like, but it's drinkable, and the emails aren't nearly as soul-crushing as they usually are. And then, suddenly, your computer freezes.

At first, you think it's just doing that thing computers do—taking a brief, unscheduled holiday because you dared to ask it to open a second tab. But then a message appears, bright and ominous: *"Your files are encrypted. Pay us ransom or lose everything."* And just like that, your morning goes from mildly optimistic to deeply existential in record time.

Cyberattacks, you see, are not the exclusive problem of sprawling multinational corporations with far too many acronyms in their names. No, hackers are equal-opportunity menaces. In fact, they seem to have developed a particular fondness for small businesses, largely because we're easy targets. Our defenses are less "high-tech fortress" and more "a lock that's slightly sticky if you jiggle it just right."

And these attacks? They're devastating. First, there's the financial hit, which is never pleasant and always arrives with the kind of unerring efficiency you wish your broadband provider had. Then there's the reputational damage, the kind that sticks around long enough to make everyone in town whisper about your "data breach" like it's a contagious rash. And finally, there's the legal trouble, which involves paperwork that is so tedious that it's likely classified as a medieval punishment in some countries.

The worst part is that so many small business owners convince themselves it could never happen to them. "Hackers? Target *me*? Why would they bother?" you might think as if cybercriminals are diligently researching your annual turnover before deciding whether you're worth the effort. Spoiler: they're not. If you've got data, they want it. If you've got weak passwords, they'll find them. And if you've ever clicked "remind me later" on a software update, they're already halfway into your system.

But don't panic! (Well, maybe panic a little—it's good cardio.) Protecting your business doesn't require you to transform into some sort of cybersecurity guru who speaks in acronyms and owns several pocket protectors. This book is not here to drown you in jargon or suggest that the only way forward is to spend a small fortune on impenetrable firewalls with unnecessarily aggressive names. Instead, it's here to offer practical, straightforward advice. Think of it as a guide to making your business a less appealing target—like putting a lock on your door that says, "Go bother someone else."

By the time you're done, you'll know how to make informed decisions, spot the warning signs of trouble, and confidently ask the IT consultant charging you an absurd hourly rate, "Do I *really* need this?" Because while you may not be able to stop every cyberattack in its tracks, you can at least make sure that the next time a hacker comes knocking, they find your door firmly shut—and maybe even booby-trapped for good measure.

The Real Costs and Risks of Cyber Attacks

Small businesses and cybersecurity breaches go together about as well as toddlers and glitter: the aftermath is expensive, messy, and impossible to fully clean up. When a breach hits, the financial damage isn't just a slap on the wrist—it's a full-on smack to the wallet. There's the ransom (because apparently hackers think you're running a Fortune 500 company), the compliance fines (fun fact: the government loves paperwork), and the cost of fixing your system (which feels like hiring a locksmith after leaving your keys in a cab). It all adds up faster than you can say, "Wait, how much?"

But it doesn't stop there—oh no. The financial hit is just the opening act. The real show comes with the damage to your reputation. A single breach can send customer trust running for the hills, clutching its metaphorical pearls. And good luck coaxing it back. Customers these days are hyper-aware of data privacy, and once your business name is uttered in the same sentence as "cyber incident," they'll be about as willing to return as someone who's just discovered their favorite restaurant has a "C" health rating.

And then there's the legal stuff, which is the corporate equivalent of having to call your mom and explain how you managed to lose her Tupperware again. Many industries have regulations that make it very clear they expect you to keep your customers' data safe. Fail to do so, and you could find yourself neck-deep in lawsuits, or worse, sending out those cheerful "We regret to inform you..." emails to every affected customer. Nothing says, "We're totally on top of things," like notifying people their data has been compromised while desperately Googling what "compliance" even means.

The point is that cybersecurity isn't just about keeping hackers out of your system. It's about protecting the future of your business and holding onto the trust of the people who rely on you. Because without trust, you're just a business with a sign that says, "We're sorry for the inconvenience" and a lot of questions about where things went wrong. So take cybersecurity seriously—or at the very least, seriously enough that hackers decide you're not worth the effort. Trust me, your future self (and your wallet) will thank you.

A Practical, Manageable Approach

We get it—cybersecurity can feel overwhelming, especially when you're juggling so many roles as a small business owner. Between managing finances, marketing, customer service, and day-to-day operations, it's easy to push cybersecurity to the bottom of the priority list. That's why this book is designed to make cybersecurity approachable and manageable. You'll find straightforward advice, checklists, practical tips, and tools to help you secure your business without getting bogged down in complexity.

Think of this book as your "CliffsNotes" for cybersecurity—a guide that breaks down the essentials into manageable steps. You'll learn the basics of protecting your business, enough to make informed decisions and recognize when it's time to call in the experts. This isn't about adding more to your already busy plate; it's about providing you with a foundation so you can feel confident and prepared.

Empowering You and Your Team

Cybersecurity isn't just the business owner's responsibility—it's a team effort. Every member of your team plays a role in keeping the business safe. Throughout this book, you'll learn how to cultivate a culture of cyber awareness within your organization, empowering your employees to spot potential threats and make smart decisions. Creating good cybersecurity habits across your team isn't about instilling fear; it's about building awareness and vigilance that make your business stronger from the inside out.

Real-World Examples and Step-by-Step Guidance

We know that theory only goes so far, which is why this book is filled with real-world examples and case studies to show you how cybersecurity issues impact businesses like yours. You'll see how other small businesses have faced and overcome these challenges, helping you understand why these concepts matter and how they apply to your operations. We'll provide step-by-step guidance on implementing cybersecurity measures, so you can take action confidently, knowing you're making smart, protective choices.

A Future-Ready Mindset

In a digital age, cybersecurity is as fundamental as locking your doors at night. Adopting a proactive, future-ready mindset can help you avoid costly mistakes down the road. Taking small, consistent steps today can strengthen your business's resilience and make it more adaptable to evolving cyber threats. This book is here to equip you with the foundational knowledge you need to secure what you've built and to be prepared for whatever comes next.

Let's Begin

As you flip through the pages of this book, congratulations are in order—you've already taken the first step toward ensuring your business doesn't end up as the plot of a cautionary tech tale. By the time you've finished, you'll not only know the basics of cybersecurity but also feel like the confident captain of your digital ship, ready to fend off cyber-pirates (metaphorical eye patches optional). You'll even be prepared to navigate the occasionally baffling world of cybersecurity professionals, who often seem to speak their own secret language. (Hint: it's mostly acronyms and exasperated sighs.)

Look, I know "cybersecurity" doesn't exactly scream "thrilling page-turner," but stick with me. I promise to keep it as engaging as possible—and maybe even a little fun. Yes, fun! After all, protecting your business from digital disaster should feel empowering, not like cramming for a midterm on a subject you didn't sign up for.

So, let's dive in together, build a rock-solid foundation, and make sure everything you've worked so hard to create stays safe, secure, and blissfully out of reach of hackers with bad intentions and worse passwords.

CYBERSECURITY 101

WHY SHOULD YOU CARE?

Introduction: The Digital Wild West

Welcome to the World of Cybersecurity

Picture this: you're a plucky entrepreneur in the Wild West, staking your claim in a lively, chaotic frontier town. There's promise everywhere—bustling streets, fortunes waiting to be made, and perhaps a saloon or two where people are indeed discussing your rising fame. But, as with all good frontier tales, danger lurks just out of sight. Bandits and outlaws are skulking around, eager to relieve you of your hard-earned gains, perhaps with a devilish grin and a six-shooter.

Fast-forward to today, and swap dusty streets for fiber-optic cables, and voilà! You're now in the "Digital Wild West." The bandits? Cybercriminals. Their six-shooters? Sophisticated software and ingenious scams. And instead of sauntering into town in broad daylight, they hide behind keyboards, plotting attacks from anywhere in the world—likely while sipping coffee, which somehow makes it feel even more unfair.

But here's the real twist: they're not just targeting big banks with sprawling headquarters and vaults guarded by security systems that look like they came from a Bond movie. Oh no, these modern outlaws have figured out that small businesses often have their "digital gold" lying about without much protection. Why go to the effort of cracking a heavily fortified safe when a cozy little shop down the road—digital speaking—has left the back door wide open and the valuables in plain sight?

Welcome, my friend, to the thrilling, slightly nerve-wracking world of cybersecurity. It's a place where opportunity and risk collide, and the stakes are nothing less than your digital livelihood. Fortunately, this isn't a tale where you're left defenseless. There are plenty of tools, strategies, and a few trusty sidekicks (AI tools and savvy practices) to help you stand tall, like a true frontier hero, against the cyber villains of our age.

Why Does This Matter to You?

If you're sitting there thinking, "I'm just a small business—cyberattacks are for the BIG GUYS," let me stop you right there. That's the cybersecurity equivalent of leaving your front door open with a neon sign that says, *"No Alarm System. Help Yourself!"* The truth is, every business—yes, even yours—has something a cybercriminal wants: customer data, financial records, intellectual property, or maybe just a chance to cause chaos because they've had a bad day.

Here's the kicker: small businesses aren't just on the menu; they're the *appetizer*. Hackers love small businesses because they're often perceived as, how shall we put it, *digitally defenseless*. With limited resources and expertise, small businesses can look like low-hanging fruit to cybercriminals—a quick, easy payday without all the hassle of taking on a corporate giant with an IT army.

Our mission here is simple: to convince you that cybersecurity is just as non-negotiable as locking up your storefront at night. You wouldn't leave your shop wide open for any passerby to waltz in and help themselves to the cash register, would you? Well, the same logic applies to your digital assets. Cybersecurity isn't just for the "big guys"—it's for anyone who values their livelihood. And spoiler alert: that includes you.

Debunking the "It Won't Happen to Me" Mindset

One of the most dangerous mindsets a small business owner can have is, "It won't happen to me." Cybersecurity threats often feel like something that only happens to someone else—until it happens to you. This complacency is precisely what cybercriminals rely on. They know that many small businesses think they're too small to be a target, which means these businesses often leave themselves exposed. This mindset can be costly, as underestimating the risk can prevent business owners from taking the steps needed to protect themselves. Remember, you don't have to be a massive corporation to catch a cybercriminal's eye; you just need to be vulnerable.

The Real Costs of a Cyber Attack

It's tempting to dismiss cybersecurity as an "IT problem"—the kind of thing you shove over to the tech folks with a shrug and a cheerful, *"Good luck!"* For small businesses, that might mean outsourcing it entirely, assuming someone else will wave their digital magic wand and make all the bad guys go away. But here's the inconvenient truth: cybersecurity isn't just about technology. It's a business problem, and a big one at that.

A cyberattack doesn't just crash your computers—it crashes your business. It snarls up your finances, throws a wrench into your operations, and leaves your customers wondering if trusting you was such a great idea after all. The fallout can be like dropping a pebble in a pond, except the ripples are less "serene lakeside meditation" and more "catastrophic tidal wave."

And let's not sugarcoat it: when an attack hits, it's not just about *"fixing the computers."* It's about recovering critical data, soothing frazzled customers, and sometimes even managing a public relations circus. In today's hyper-connected world, a breach isn't just an IT headache—it's a full-blown business migraine that demands attention from the top down. So, while your IT team (or outsourced superhero) might be your front line, make no mistake—this battle belongs to everyone.

Breaking Down the Dollars and Cents

Let's crunch some numbers, shall we? Cyberattacks aren't just costly—they're the financial equivalent of setting your wallet on fire and then realizing your insurance doesn't cover *"acts of hackers."* It's not just about the immediate hit to your bank account; it's the long-term fallout that really stings. Here's a glimpse of the damage these digital disasters can wreak:

Direct Costs:

Fines and Penalties: If your business handles sensitive customer information, like credit card details or personal health information, you're legally required to protect that data. A breach can lead to regulatory fines, especially if you're non-compliant with regulations like GDPR or HIPAA. For a small business, these fines can be substantial enough to put you at serious financial risk.

Lost Revenue: Imagine having to shut down operations for a day, a week, or even longer to deal with a cyber incident. During that downtime, your business isn't making money, but bills are still piling up. The longer your business is offline, the greater the impact on your bottom line.

Recovery Costs: After a cyber attack, businesses often need to pay for data restoration, cybersecurity professionals, and enhanced security measures to prevent future attacks. Recovery costs can add up quickly, especially if your data was corrupted or encrypted by ransomware.

Indirect Costs:

Reputational Damage: In the age of social media and online reviews, word spreads fast. A cyber attack can damage your business's reputation and erode trust among customers. If customers feel their personal information isn't safe with you, they're likely to take their business elsewhere.

Customer Attrition: For many small businesses, loyalty is built on trust. A data breach can shake that trust, causing customers to leave and seek out competitors. Rebuilding that trust is an uphill battle, and some customers may never return.

Increased Insurance Costs: If you have cyber insurance, a breach might lead to increased premiums. Some insurers might view your business as a higher risk after a breach, raising costs and making insurance harder to afford.

Real-Life Example of Financial Impact on a Small Business

In a mid-sized law firm celebrated for its legal expertise, a team of accomplished attorneys, thrived on delivering top-tier legal services to both individuals and businesses. With a reputation that spanned across diverse practice areas, the firm seemed unshakable in its standing. However, behind its polished exterior, the firm harbored a critical blind spot: a complete lack of a cybersecurity plan.

The turning point came when the firm publicly announced a major development—one of its partners was leaving to join another prestigious law firm. The news attracted widespread attention, as it was shared across media outlets and became a topic of discussion among peers and clients. Amid the congratulations and inquiries, an inconspicuous phishing email landed in the inbox of one of the firm's partners. Disguised as an urgent request from a local bank, it asked for an update to account information.

Believing the email to be legitimate, the partner forwarded it to their paralegal to handle. Acting promptly, the paralegal clicked the link and landed on what appeared to be a Microsoft Online login page. Without questioning its authenticity, they entered their credentials—unknowingly handing access over to a hacker.

In mere moments, the hacker gained control of the paralegal's email account. Wasting no time, they used the compromised account to send the same phishing email to over 300 contacts, triggering chaos. Clients and colleagues began flooding the firm's phones, alerting them to suspicious emails being sent from the paralegal's account.

The firm quickly contacted their local IT support company, initiating an incident response to address the breach. IT specialists secured the compromised accounts and began an investigation. But the phishing attack, as disruptive as it was, turned out to be just a decoy. The hacker's real objective lay elsewhere.

While the firm scrambled to contain the fallout, the hacker turned their attention to the firm's accountant, who was in the midst of managing financial transactions related to the departing partner's merger with the other law firm. Using the access gained from earlier phishing activities, the hacker intercepted email communications about a wire transfer. Posing as a representative from the firm's bank, they sent fraudulent account details to the accountant.

Unaware of the deception, the accountant forwarded the account details to the other law firm, which subsequently wired $10,000 to the fraudulent account. It wasn't until weeks later, during routine financial reconciliation, that the error was discovered. A panicked back-and-forth with the other firm confirmed the worst—the money was gone.

The IT investigation revealed the full scale of the breach. The hacker had not only intercepted communications but had also set up email forwarding rules in the accountant's account, ensuring that sensitive emails were copied to an external address. The firm faced devastating consequences: $10,000 lost to the fraudulent transaction, an additional $10,000 spent on emergency IT services, and immeasurable damage to its reputation and client trust.

In total, the cyberattack cost the firm well over $25,000 by the time all was said and done. Yet, the financial loss was just part of the story. The breach exposed how a lack of cybersecurity protocols left the firm vulnerable to exploitation—a hard-learned lesson that no organization, no matter its size or prestige, can afford to ignore.

In Summary

In this digital age, cybersecurity is as fundamental as locking up your storefront at night. Cyber attacks aren't just a possibility—they're a real and present threat to businesses of all sizes, and small businesses are no exception. The costs of a cyber attack extend beyond lost data or downtime; they affect your reputation, your customer relationships, and, ultimately, your bottom line.

This chapter aims to show you why cybersecurity should be a priority, not an afterthought. By understanding the risks and costs associated with cyber attacks, you're already taking the first step toward safeguarding your business. Remember, a cyber-attack is not an "if" but a "when" for many small businesses. Preparing for that "when" could mean the difference between bouncing back stronger and facing a crisis that could put your business's future in jeopardy.

Cybersecurity doesn't have to be overwhelming or overly technical. Throughout this book, we'll equip you with the essential knowledge and tools to protect your business without straying from what you do best. This journey starts with recognizing the importance of cybersecurity—because, in the Digital Wild West, every small business needs a strong defense to keep their digital gold safe.

Why Small Businesses are Prime Targets

The Myth of "Only Big Companies Get Hacked"

When most people think of cyber attacks, the mental image is practically cinematic: a shadowy figure in a dark hoodie, hunched over a keyboard, illuminated only by the glow of their monitor as they wage a digital war against massive corporations or government agencies. It's easy to imagine hackers aiming their virtual slingshots at tech giants, billion-dollar banks, or sprawling conglomerates with vast reserves of data and deep pockets. Small business owners, on the other hand, tend to think, *"Why would anyone waste their time on my little shop?"*

It's a comforting thought, really. Unfortunately, it's also nonsense.

Cybercriminals are not exclusively after the big fish. In fact, they often prefer to cast their nets in waters teeming with smaller prey. Why? Because small businesses, bless them, are like houses with open windows and keys hidden under doormats labeled "Key." While big corporations might have fortress-like cybersecurity systems, small businesses are often delightfully—at least from a hacker's perspective—unprotected.

The idea that "only big companies get hacked" is undoubtedly one of the most persistent and harmful myths out there. It lulls small business owners into a false sense of security, leaving them blissfully unaware that their digital doors are wide open.

For a hacker, small businesses are the equivalent of a convenience store with the lights on but no cashier behind the counter. They don't need weeks to breach a maze of firewalls or decrypt layers of high-tech security. They just waltz in, grab the data—customer records, credit card numbers, login credentials—and waltz right back out again.

The notion that small businesses are "too small to hack" isn't just wrong—it's dangerous. It's the kind of thinking that hackers count on, and sadly, it's why so many small businesses find themselves blind-sided when an attack occurs. The truth is simple: if your business has anything of value—and spoiler alert, it does—you're on their radar. It's not a question of *if* they'll target you but *when*.

Why Hackers Love Small Businesses

If cybercriminals were hunters, small businesses would be their favorite watering hole—easy to spot, easy to approach, and often blissfully unaware of the danger. Hackers aren't random in their targets; they're strategic, calculating, and, let's face it, rather annoyingly clever. They go after what's easiest to crack and most likely to pay off. For many, small businesses are the perfect jackpot. Here's why:

Less Investment in Security

Small businesses often operate on tight budgets, and cybersecurity tends to end up on the "nice to have" rather than the "must have" list. Every dollar spent on security is a dollar not going toward things like product development, marketing, or keeping the lights on, so many small businesses make do with the digital equivalent of a flimsy padlock. Maybe they've got basic antivirus software, or perhaps they've downloaded a free security tool after reading something scary online.

Hackers, of course, are keenly aware of this. They know small businesses aren't shelling out for the sophisticated, multi-layered defenses that big corporations employ. For them, it's like strolling past a house with no alarm system and an "Open for Business" sign in the front yard. The path of least resistance is irresistible, and small businesses often provide exactly that.

Supply Chain Vulnerabilities

Now, if you're a small business thinking, "But I don't have anything they'd want!"—think again. Hackers aren't just after you; they're after who you're connected to. Small businesses play a critical role in supply chains, serving as the invisible glue that holds larger operations together. And hackers love a good backdoor.

Picture this: you're a modest parts supplier for a major car manufacturer. To you, it's just another day of shipping bolts and widgets, but to a hacker, you're a side door to the big leagues. Breach your systems, and they've got a foothold in the car company's network. It's like sneaking into a VIP party by charming the doorman. Small businesses often don't realize that their connections to larger organizations make them prime stepping stones for attackers looking to infiltrate the supply chain.

More Human Error

Ah, human error—the gift that keeps on giving for cybercriminals. Small businesses often run on lean teams where everyone wears multiple hats, from marketing wizards to IT troubleshooters. Unfortunately, this "jack of all trades" setup means there's usually no one dedicated to cybersecurity, let alone providing regular training or keeping up with best practices.

Hackers thrive on this. Employees using "Password123" for everything? Jackpot. Is someone clicking on a phishing email that promises a free gift card? Even better. The smaller the team, the more likely someone's going to slip up, and hackers know it. It's not that small businesses are careless; it's that they're stretched thin. And that, dear reader, makes them a prime target.

Hackers love small businesses because they're easy, unguarded, and often connected to bigger prizes. It's not personal—it's just business. But that's exactly why small businesses can't afford to ignore cybersecurity. Because for hackers, there's nothing small about the opportunity you represent.

Numbers Don't Lie: Stats on Small Business Cyber Attacks

If you're still not convinced that small businesses are prime targets, let's look at the numbers. Statistics show that small businesses are under constant threat from cyberattacks, and the numbers are rising every year.

For instance, recent studies have shown that nearly 43% of all cyberattacks target small businesses. Think about that for a moment. Almost half of all cyberattacks are aimed at businesses with fewer than 100 employees. This isn't a fluke or a random coincidence; it's a clear indication that hackers are actively seeking out small businesses. And it doesn't stop there. Studies also reveal that over 60% of small businesses hit by a cyberattack end up closing their doors within six months. The financial and reputational impact of an attack can be crippling, especially for businesses with limited resources to recover.

Furthermore, surveys show that around 70% of small businesses are unprepared for a cyberattack. They lack formal policies, don't conduct regular security assessments, and often don't have a dedicated IT team to manage threats. Hackers know this. They understand that small businesses are often unprepared, and they take advantage of that vulnerability.

A Story of a Small Business Targeted by Hackers

To make this a little more real, let's look at a hypothetical (but all too realistic) example. Picture a small family-owned bakery. It's a cozy place, well-loved in the community, where customers come for freshly baked goods and friendly service. The bakery owner has a website for online orders and uses a basic point-of-sale system to manage transactions. Everything seems to be running smoothly—until one day, it all comes crashing down.

One of the bakery's employees receives an email that looks like it's from their supplier, asking them to verify an order by clicking a link. Without thinking twice, the employee clicks on the link and unwittingly downloads malware onto the bakery's system. Within minutes, the malware has locked up their point-of-sale system, encrypting all their files. The hackers send a ransom demand: pay up, or lose access to everything.

The bakery owner panics. All their records are frozen, and they can't process any transactions. They're losing hundreds of dollars in sales each day, and the stress is mounting. On top of that, the hackers managed to access customer payment information stored in their system. Now, the bakery is facing not only a financial crisis but also a loss of customer trust. Some loyal customers are angry that their data was compromised, and a few have even decided to take their business elsewhere.

In the end, the bakery owner pays the ransom, but the financial impact doesn't stop there. They have to hire a cybersecurity consultant to clean up their system and restore what they can. They also face higher insurance premiums due to the breach, and they're left wondering if they'll ever regain their customers' trust. This small, seemingly insignificant bakery has become yet another victim of cybercrime—a stark reminder that no business is too small to be targeted.

Types of Threats in Plain English

Phishing: The Digital Bait-and-Switch

Imagine you're fishing on a quiet lake. You toss your line in the water with a juicy piece of bait, hoping that an unsuspecting fish will swim by and bite. Now, think of this in digital terms. Hackers are the ones doing the fishing—or rather, "phishing"—and the bait is an email or text message that looks convincing but is designed to trick you.

Phishing is the classic "bait and switch," but instead of a hook, it uses cleverly disguised messages to lure people into sharing sensitive information. Here's how it works: a hacker crafts an email that looks like it's from a trusted source—maybe your bank, a popular online store, or even a company executive. The email might say something urgent, like, "Your account has been compromised. Click here to reset your password," or "There's an issue with your latest invoice. Please confirm your payment details." It feels legitimate, especially when there's a sense of urgency attached. They want you to act fast, without thinking things through.

The moment an employee clicks the link or enters sensitive information, they've taken the bait. Instead of resolving an issue, they unknowingly handed over their login credentials, financial data, or other confidential information directly to the hacker. Phishing attacks are common because they are effective. It's a low-effort, high-reward tactic for cybercriminals, and it can be devastating for businesses of any size. All it takes is one person on your team to fall for a phishing scam, and suddenly, hackers have the keys to your digital kingdom.

In this era, phishing isn't limited to emails. Hackers also use text messages (called "smishing") and even voice calls ("vishing") to pull off these scams. The best defense? Awareness. By educating your team about phishing tactics and how to spot suspicious messages, you can avoid falling for these digital bait-and-switch schemes and protect your business from costly breaches.

Ransomware: Holding Your Data Hostage

Imagine waking up one morning, heading into your business, and finding that all your files are locked up tight—customer records, financial spreadsheets, inventory lists, everything. Then, a message appears on your screen, something like, "We've taken your data hostage. Pay us, or you'll never see it again." This isn't the plot of a thriller; it's ransomware in action—a "digital kidnapping" that's happening to businesses every day.

Ransomware is a type of malicious software that cybercriminals use to gain control of your data. Once they're in, they encrypt your files, making them completely inaccessible to you. Then they hold your data hostage, demanding a ransom—usually in cryptocurrency—for the decryption key that will supposedly unlock everything. Think of it as a ransom note left by a kidnapper, but instead of a person, they've taken your valuable data.

This type of attack typically starts with a seemingly innocent action, like an employee clicking on a malicious link in an email or downloading an infected attachment. Once ransomware is installed, it spreads quickly, locking files and even entire systems. In some cases, hackers give you a ticking clock to add to the pressure: "Pay within

72 hours, or your data will be deleted forever." It's a brutal tactic, designed to push businesses into a corner and force them to cough up the ransom.

Why is ransomware one of the fastest-growing cyber threats? Simply put, it's profitable. Cybercriminals know that for many businesses, paying the ransom might seem like the quickest way to get back to normal operations. In fact, some hackers even provide "customer support" to ensure that victims know how to make the payment. And as if that weren't bad enough, paying the ransom doesn't always guarantee you'll get your data back—some hackers simply take the money and run.

For small businesses, a ransomware attack can be especially devastating. Paying the ransom is one option (if you're willing to take that risk), but there are also costs associated with downtime, data recovery, and potential reputational damage. Imagine not being able to access your data for days or weeks. For many small businesses, that kind of disruption is hard to recover from.

The best defense against ransomware is to stay vigilant, keep your systems updated, and, above all, have regular backups of your data. If your files are backed up securely, you'll be in a much better position to recover without having to negotiate with cybercriminals. Remember, ransomware isn't just a tech problem; it's a business problem that affects every part of your operation. Staying prepared can make all the difference in a world where digital kidnappings are, unfortunately, becoming all too common.

Malware: Viruses, Trojans, and Worms, Oh My!

Think of malware as the all-purpose bad guy of the digital world. It's not one specific type of villain; it's more like an entire gang of troublemakers, each with its own nasty tricks. At its core, "malware" is just a catch-all term for any kind of software designed to cause harm, steal information, or generally make your digital life miserable.

Imagine malware as a shady character sneaking around your business, rifling through your files, leaving doors open for other crooks, or even planting time bombs that could go off at any moment. It's a sneaky umbrella term that covers a variety of cyber threats, including:

- **Viruses**: These are the classic troublemakers. Just like a flu virus spreads from person to person, a computer virus spreads from file to file or system to system, often causing havoc as it goes. Viruses are notorious for slowing down systems, corrupting files, and generally making a mess of things.

- **Trojans**: Remember the story of the Trojan Horse? The Greeks left a giant wooden horse outside the gates of Troy, and the Trojans (thinking it was a gift) brought it inside. Little did they know, Greek soldiers were hidden inside, waiting to attack. In the digital world, a Trojan works the same way. It's a program that looks harmless—like a free game or a helpful app—but once you let it in, it can open the door for hackers to access your system and steal your data.

- **Worms**: Worms are like viruses' overly ambitious cousins. They don't just stick to one system—they want to get out and explore! Worms spread rapidly through networks, finding their way onto as many computers as they can reach. They often don't even need a "host" file to attach to; they burrow in and keep multiplying, slowing down networks and eating up resources.

Malware is the digital equivalent of finding an unwanted guest wreaking havoc in your house. It sneaks in, messes things up, and sometimes invites even more bad actors to join the party. And while the tech world has plenty of solutions to combat these pests, the best first line of defense is awareness. By understanding what malware is (in simple, non-technical terms), you're already one step ahead in keeping your business safe from these digital troublemakers.

Social Engineering: Psychological Manipulation

Imagine you're in a detective novel where the villain doesn't just break into safes or hack computers—instead, they rely on smooth-talking, clever disguises, and psychological tricks to get what they want. That's social engineering in the cyber world. It's less about hacking computers and more about hacking people. In simple terms, social engineering is the hacker's way of "playing mind games" to trick people into giving away valuable information.

Hackers who use social engineering are masters of psychological manipulation. They might not even need technical skills because they've learned how to exploit basic human tendencies like trust, curiosity, or even fear. Their goal? To get someone to reveal sensitive information, grant access to restricted areas, or unwittingly perform actions that compromise security. Let's look at a couple of their favorite tricks:

- **Pretexting**: In a pretexting scenario, the hacker creates a convincing backstory—or "pretext"—to gain someone's trust. Imagine receiving a call from someone who says they're from your bank's fraud department. They sound professional and knowledgeable, and maybe even drop a few personal details about your account to build credibility. Then, they ask you to "verify" your password or PIN. In reality, this person isn't from your bank at all; they're just playing a role to get you to spill sensitive information. By the time you realize it, they're already halfway into your account.

- **Impersonation**: This is the classic "wolf in sheep's clothing" approach. Hackers might impersonate a trusted figure within your company, like a manager or even the CEO, sending an urgent email to employees. The email might say something like, "Hey, I need you to wire $10,000 to this account ASAP for a new vendor. I'll explain later—just make it happen." People naturally want to follow instructions from authority figures, so they comply without questioning. But that "urgent" email was from a hacker, not the CEO, and now $10,000 is gone.

Social engineering is dangerous because it bypasses all the firewalls, passwords, and fancy security software by targeting the most vulnerable part of any organization: human nature. Hackers know that even the most secure systems can be compromised with the right words and a little bit of trickery. And for small businesses, where everyone wears multiple hats and cybersecurity training might be minimal, these mind games can be especially effective.

The best defense against social engineering is awareness. By understanding these tactics, you and your team can recognize when someone's trying to play mind games and stay one step ahead of the manipulators. Remember, in the world of cybersecurity, trust should always come with a healthy dose of skepticism!

Business Email Compromise (BEC): The CEO Scam

Imagine this: you're working away at your desk when an email pops up from your company's CEO. It's marked as "urgent" and seems to be coming straight from the top. The message is short but direct: "Can you wire $15,000 to this new vendor immediately? We're on a tight deadline. Don't tell anyone else—just handle it quickly, please."

You pause for a second but then think, "Well, it's from the CEO, so it must be legit." You follow the instructions, send the money, and go back to your day. But a few hours later, you get a call—no one at the company authorized that payment. The CEO never sent that email. You've just been hit by a Business Email Compromise (BEC) scam.

What is BEC? BEC scams, sometimes known as "the CEO scam" or "executive impersonation," are a sophisticated form of phishing where attackers specifically target businesses by pretending to be someone with authority, usually an executive or high-ranking employee. These attacks are carefully crafted and rely on the target's trust in their colleagues and leaders.

Hackers use various tricks to make these emails look genuine. They might spoof the CEO's email address to make it appear legitimate, or they could compromise an actual employee's email account to send the message. They study the company, learn the structure, and wait for the right moment to strike. Once they send the message, they count on the recipient's willingness to act quickly—often bypassing normal checks and balances because, well, it's "urgent" and from the boss.

How Do BEC Scams Work? Here's how attackers pull off these scams:

1. **Research and Recon**: Hackers do their homework. They look up information about the company's executives, organizational structure, vendors, and even ongoing projects. Social media, LinkedIn, and company websites often provide them with all the details they need.

2. **Spoofing or Compromising Accounts**: Attackers either spoof an email address to make it look like it's coming from a trusted source or, worse, gain access to an actual company email account. The latter is more effective because it's coming from a genuine address, so there's little reason for anyone to doubt its authenticity.

3. **Crafting the Message**: The message is usually short, urgent, and to the point. It often includes phrases like, "I need you to handle this ASAP," "This is confidential—don't discuss it with anyone," or "We're on a tight deadline." The attacker's goal is to create a sense of urgency that prevents the recipient from thinking things through or checking with others.

4. **Requesting Money or Sensitive Information**: The email might instruct the recipient to wire money to a specific account, pay a new vendor, or share sensitive company information. Since it appears to be coming from a trusted executive, employees may feel pressured to act quickly without double-checking the request.

Why Are BEC Scams So Effective? BEC scams work because they exploit trust, authority, and urgency. When an email appears to be from the CEO or another senior executive, most employees are inclined to follow instructions without hesitation. After all, questioning the boss is not something most people do, especially in a high-pressure situation.

Additionally, BEC scams don't rely on malware or suspicious links, which means they can slip past standard security filters. It's just a simple email with straightforward instructions, but that simplicity is what makes it so dangerous.

Protecting Your Business from BEC Scams The best defense against BEC scams is a mix of awareness, verification procedures, and cautious skepticism. Here are a few basic practices to protect against this type of fraud:

1. **Always Verify**: If you receive an urgent request for money or sensitive information, especially one that deviates from normal protocol, verify it through another channel. A quick phone call to the supposed sender can confirm whether the request is legitimate.

2. **Establish Protocols**: Create and enforce clear policies for financial transactions, requiring multiple layers of approval, especially for large sums. Make it a rule that any unusual or high-value requests need a second opinion, even if they appear to come from someone senior.

3. **Training and Awareness**: Educate employees about BEC scams and encourage them to question unusual requests. Make sure everyone understands that it's okay—and encouraged—to verify unusual requests, even if they come from the boss.

4. **Use Email Security Tools**: Some email security solutions can help detect spoofed addresses or identify suspicious patterns in emails. While no tool is foolproof, this added layer of protection can sometimes catch red flags that employees might overlook.

In a world where trust is often exploited, it pays to be a little skeptical. Business Email Compromise scams leverage authority and urgency to trick employees into acting against their better judgment. By understanding how BEC works, businesses can put up defenses to stop these "CEO scams" before they cause financial damage. Remember: even if it's "from the CEO," a little extra verification can go a long way in protecting your business.

Insider Threats: When the Danger Comes from Within

When we think of cybersecurity threats, we usually picture shadowy hackers lurking on the internet, trying to break into our systems from a distance. But sometimes, the danger isn't coming from some faraway attacker. Instead, it's coming from someone inside the building. This is what we call an "insider threat," and it's a reminder that not all risks are external.

An insider threat doesn't necessarily mean you've got a spy or a saboteur in your midst (although, in rare cases, that can happen). It could be something as simple as an employee who clicked on the wrong link, accidentally exposing the entire network to malware. Or maybe it's someone who left the company on bad terms and still has access to sensitive data. The point is that insider threats can take many forms, from innocent mistakes to intentional harm, and they're worth paying attention to.

Here are some common examples of insider threats:

- **The Disgruntled Employee**: Sometimes, an unhappy employee who feels mistreated or unappreciated may decide to take "revenge" on the company. This could mean anything from leaking sensitive information to deleting critical files or even sabotaging systems. While this is rare, it's a real risk, especially if former employees still have access to company accounts or data after they leave.

- **The Careless Clicker**: Not all insider threats are intentional. In many cases, an employee might simply make a mistake—like clicking on a phishing email, using a weak password, or downloading a file from an unsafe source. These innocent actions can have serious consequences, potentially exposing the entire organization to cyberattacks. Hackers love this type of insider threat because it's easy to exploit: all it takes is one untrained employee to open the door.

- **The Overly Helpful Staff Member**: Picture this—someone calls your office pretending to be an IT technician and asks an employee for their login credentials. Some employees, wanting to be helpful, might hand over the information without thinking twice. This kind of insider threat, where someone unknowingly gives access to a bad actor, is called "social engineering." It's a reminder that even the best-intentioned employees can sometimes create security risks.

- **The Ex-Employee Who Still Has Access**: Sometimes, companies forget to revoke access when an employee leaves. This can lead to big problems if that former employee decides to use their old credentials to access company systems. Even if they have no malicious intent, their login information could be compromised or sold to hackers on the dark web.

So, What Can You Do About Insider Threats?

Dealing with insider threats doesn't mean you need to be paranoid or start suspecting everyone around you. It's simply about taking a few sensible steps to reduce risk:

1. **Implement Access Controls**: Not every employee needs access to every piece of information. By limiting access based on roles and responsibilities, you reduce the chances of an insider threat—whether accidental or intentional. For example, the accounting team shouldn't have access to sensitive HR records, and vice versa.

2. **Train Your Team on Cyber Awareness**: Many insider threats come from simple mistakes or lack of awareness. By educating your employees on cybersecurity best practices—like recognizing phishing emails, using strong passwords, and verifying requests for sensitive information—you can cut down on accidental risks.

3. **Monitor and Log Activity**: Having systems in place to monitor and log employee activity can help detect unusual behavior. For instance, if someone suddenly starts accessing files outside of their usual scope, it might be a red flag worth investigating.

4. **Revoke Access Promptly**: When employees leave, make sure you remove their access to company systems immediately. This simple step can prevent a lot of headaches down the road.

5. **Create a Culture of Security**: Encourage a culture where employees understand the importance of cybersecurity and feel comfortable reporting suspicious activity—whether it's

something they see internally or an external phishing attempt. Emphasize that cybersecurity is everyone's responsibility, and a strong security culture can go a long way.

The Bottom Line: Insider threats don't mean you have to look over your shoulder every time you walk into the office. It's not about distrusting your team but rather about recognizing that sometimes, mistakes happen. By putting sensible controls in place and fostering a security-aware culture, you can minimize these risks. Remember, not all threats come from outside the company. By being aware of potential insider threats, you're taking another step toward keeping your business safe from within.

The First Steps to Take

Step 1: Take Stock of Your Digital Assets

Imagine you're moving to a new office. Before you pack up, you'd probably take a quick inventory of everything valuable—computers, files, coffee maker (can't forget that), and all the essentials that keep your business running smoothly. Cybersecurity is no different. Before you can protect what matters, you need to know exactly what you have that's worth protecting.

The first step in securing your business is to make a list of your digital assets. Think of it as a "cyber inventory." These assets are the digital equivalents of your most valuable physical items, but in some cases, they can be even more important. Here are some key categories you should consider:

- **Customer Data**: This could include anything from email

addresses and phone numbers to billing information and purchase history. If your business stores customer details, this is one of your most critical assets. Not only is it valuable to you, but it's also attractive to hackers who can use it for identity theft or sell it on the black market.

- **Financial Information**: Your accounting records, payment processing data, and banking information are all part of this category. Financial information is a prime target for cyber-criminals because it's directly linked to your money. Imagine the chaos if this information were compromised or inaccessible.

- **Intellectual Property**: Do you have proprietary designs, formulas, or business plans that give you a competitive edge? Your intellectual property is unique to your business and is often what makes you stand out in the market. Losing it or having it leaked could harm your reputation or even lead to lost revenue if competitors get their hands on it.

- **Employee Information**: From Social Security numbers to health records, if you store employee information, it's your responsibility to protect it. Data breaches that expose employee records can lead to legal troubles, not to mention the strain it places on employee trust and morale.

Knowing what you're protecting makes it much easier to focus your cybersecurity efforts where they matter most. Encourage yourself (and your team) to take a step back and really think about all the valuable information your business holds. Make a simple list or spreadsheet that documents each asset category and describes what type of information is involved.

Taking stock of your digital assets might feel like a tedious task, but it's the foundation for everything that comes next. You wouldn't leave your business's valuables out in the open; the same rule applies to your digital assets. By understanding what you have, you're taking the first, crucial step toward keeping your business safe from cyber threats.

Step 2: Identify Your Weak Spots

Once you know what you need to protect, the next step is to figure out where you might be vulnerable. Think of this as a mini self-assessment—a quick walk-through to identify any "cracks" in your cybersecurity foundation that a hacker could exploit. It's like checking your home for unlocked windows and doors before you go on vacation. You don't need a cybersecurity degree to get started here; just a little honest reflection and a keen eye for potential gaps.

Here are a few key areas to consider:

- Outdated Software: Take a look at the software and systems you use daily. Are they up-to-date? Outdated software can be a significant weak spot because it often lacks the latest security patches. Hackers know this, and they actively look for businesses running older versions of programs to exploit known vulnerabilities. If you've ever clicked "Remind me later" on a software update prompt, now might be the time to hit "Update" instead.

- Weak Passwords: This one's big. Weak passwords are like leaving your business's digital front door unlocked. Are you or your employees still using easy-to-guess passwords, like "password123" or "qwerty"? If so, it's time for a password upgrade. Ideally, you want to use complex, unique passwords for each account—something that can't be cracked by a hacker in seconds. If remembering all these passwords sounds daunting (and, trust me, it is), consider using a password manager to keep things organized and secure. With a password manager, you can easily use new passwords with each site and make sure each of them is random and extremely complicated... all the while never having to remember another password.

- Lack of Employee Training: Cybersecurity isn't just a tech issue; it's also a people issue. Your employees are the ones who interact with your systems daily, and their actions (or mistakes) can either protect or expose your business. Ask yourself: have you provided training to help them recognize phishing emails, create strong passwords, or follow basic security protocols? If the answer is no, this could be a significant weak spot. Even a little training can go a long way in preventing accidental breaches.

- Unsecured Wi-Fi and Networks: Many small businesses rely on Wi-Fi networks for daily operations. But if your network isn't properly secured, it can be an easy target for hackers. Ensure that your Wi-Fi is protected with a strong password, and avoid using public or unprotected networks to access sensitive information. Also, consider using a Virtual Private Network (VPN) to add an extra layer of security when accessing important data remotely.

- Unmonitored Access Points: Access points include any devices or systems that connect to your network, such as computers, smartphones, or even smart devices like printers. Every access point represents a potential entryway for cybercriminals. If you haven't been monitoring who and what connects to your network, it's time to start. Unmonitored or unauthorized devices can create serious security risks.

Think of this self-assessment as a checklist. Go through each of these areas and ask yourself where you might have vulnerabilities. Are you staying on top of software updates? Are passwords strong and unique? Have employees been given basic cybersecurity training? Taking a bit of time to identify your weak spots now can save you a world of trouble down the road.

And remember, identifying vulnerabilities isn't about pointing fingers or feeling bad. It's about taking a proactive step to improve your defenses. Once you know where your weak spots are, you can start working on shoring them up, making your business a much tougher target for cyber threats. This simple self-assessment is like giving your cybersecurity foundation a tune-up, making sure it's ready to withstand the storms of the digital world.

Step 3: Set Up Basic Defenses

Once you've identified your digital assets and taken stock of your weak spots, it's time to set up some basic defenses. Think of this as installing locks on your doors, setting up a home alarm system, and making sure you have a spare key in a safe place. These basic cybersecurity measures aren't complicated, but they're crucial for keeping your business secure in the digital world. Let's dive into a few simple steps that can make a huge difference in protecting your data and systems.

Password Protection: Building Strong First Lines of Defense

Passwords are your first line of defense against unauthorized access, but they're only as strong as you make them. Imagine leaving the keys to your business under the doormat—would you feel secure? Weak passwords are the digital equivalent of that. Hackers know all the common tricks and often use automated tools that can guess simple passwords in seconds. So, let's talk about what a strong password looks like and why it's worth considering a password manager.

A strong password should be:

- At least 12 characters long

- A mix of uppercase and lowercase letters, numbers, and special characters

- Unique for each account (don't reuse passwords across different platforms)

Creating and remembering these kinds of complex passwords can feel overwhelming, especially if you're juggling multiple accounts. That's where a **password manager** comes in. A password manager securely stores all your passwords in one place and generates strong, unique passwords for each account. Think of it as your digital keychain. Instead of remembering dozens of passwords, you only need to remember one master password to access the manager. It's a simple way to strengthen your defenses without adding stress.

By upgrading your password habits and possibly investing in a password manager, you're making it significantly harder for hackers to crack your accounts. Remember, passwords are your digital "locks," so make them tough to pick!

Multi-Factor Authentication (MFA): Adding an Extra Layer of Protection

Imagine you've got a super-strong lock on your door, but there's also a security guard inside checking IDs before anyone can enter. That's essentially what **Multi-Factor Authentication (MFA)** does for your accounts. It adds an extra layer of security by requiring not just a password, but also a second form of identification to verify that you're really you.

Here's how it works: when you log in, MFA might ask you to enter a code sent to your phone, or use a fingerprint scan if your device supports it. This way, even if someone manages to steal your password, they'd still need access to your second authentication method to break in. Think of MFA as a two-step process: something you know (your password) and something you have (a code sent to your device or a fingerprint). It's like double-locking your doors, but without the hassle.

Enabling MFA is easier than it sounds. Most online services and platforms offer it, and it's usually just a matter of going into your account settings and turning it on. Sure, it adds an extra step to your login process, but it's worth it for the added security. Hackers typically look for easy targets, and MFA makes your accounts significantly harder to crack, effectively turning them away in search of easier prey.

By setting up MFA, you're taking a proactive step to protect your business. This simple layer of security can prevent unauthorized access and gives you peace of mind, knowing that even if someone has your password, they still can't get in without that second form of verification.

Regular Backups: Your Safety Net When Things Go Wrong

Imagine if your business's entire customer list, financial records, and other critical data were wiped out in an instant. It's a terrifying thought, but cyberattacks like ransomware or even hardware failures can make it a reality. That's why having **regular backups** of your data is so important—it's your safety net in case things go south.

Backing up your data means creating a copy of all your important files and storing it in a secure location. If your primary data is compromised, you can restore it from this backup, minimizing downtime and helping you recover faster. Think of it as having an insurance policy for your digital assets. Here are a few key points to keep in mind:

1. **Frequency**: Back up your data regularly. For most businesses, a weekly backup schedule is sufficient, but if your work involves constantly updating files, consider daily backups. The more frequently you back up, the less data you'll lose in the event of a disaster.

2. **Location**: Store your backups in a safe place. Ideally, this should be separate from your primary systems. Consider using a cloud-based backup service, which stores your data in secure, remote servers, or an external hard drive that's kept offline when not in use. This way, if your main system is compromised, your backup remains safe.

3. **Testing**: Don't just assume your backup will work. Test it periodically to ensure that you can actually restore your data

from it if needed. There's nothing worse than finding out your backup is corrupted when you're already dealing with a crisis.

Regular backups protect you from a wide range of threats, including ransomware, accidental deletions, and hardware malfunctions. If a cyberattack does hit, having a recent backup can be the difference between a minor setback and a business-ending disaster.

Putting It All Together: The Basic Defense Trio

By setting up strong passwords, enabling MFA, and scheduling regular backups, you're creating a solid foundation for your cybersecurity efforts. Think of these steps as locking your doors, setting up a security alarm, and keeping a spare key in a safe place. They're simple, straightforward actions that can save you a world of trouble in the long run.

Cybersecurity doesn't have to be complicated, especially when you start with the basics. These small steps may seem simple, but they make a big difference in keeping your business safe. Once you have these defenses in place, you're not just protecting your business—you're also building the confidence to handle whatever the digital world throws at you. So go ahead, set up those defenses, and take the first step toward a safer business.

Step 4: Start a Culture of Cyber Awareness

Let's face it: cybersecurity isn't just a "tech problem" or something for the IT department (if you have one) to worry about. In today's digital landscape, everyone in the company plays a role in keeping the business secure. Think of cybersecurity like workplace safety—just as you wouldn't let employees ignore a slippery floor or faulty equipment, you also can't afford to ignore basic security practices. And just like workplace safety, cybersecurity works best when it becomes part of your company's culture.

Creating a culture of cyber awareness doesn't require extensive training or complex procedures. It's about instilling simple, proactive habits in every employee that will collectively make your business a tougher target for cybercriminals. Here's how you can start fostering a culture of cyber awareness, one small step at a time.

Making Cybersecurity Part of Everyday Conversations

Cybersecurity shouldn't be a "once-a-year" topic, like a fire drill or an HR seminar that everyone quickly forgets. Instead, encourage regular, casual conversations about security. These don't have to be big, intimidating discussions. Sometimes, a quick reminder in a team meeting or a friendly email notice about staying vigilant can go a long way.

For instance, send out occasional reminders like, "Hey team, remember not to click on any suspicious links or emails that look fishy!" Or, "Just a heads up, if you see a software update pop up on your screen, go ahead and install it—keeping things up-to-date helps protect us all!" These reminders may seem small, but they serve as gentle nudges to keep cybersecurity top of mind. By normalizing these discussions, you're letting your team know that security isn't just an IT thing; it's everyone's responsibility.

Creating Simple Rules Everyone Can Follow

Not everyone in your business needs to be a cybersecurity expert, but they should know a few basics. Think of this as setting "ground rules" for safe behavior online, just like you'd set guidelines for workplace etiquette. You can start with a few simple practices that are easy for everyone to understand and follow:

- **Don't Click on Suspicious Links**: Remind employees that if they receive an email from an unknown source with a link or attachment, they should be cautious. Encourage them to verify the sender or ask a colleague before clicking on anything that seems off.

- **Avoid Reusing Passwords**: Reinforce the idea that passwords should be unique for each account. Emphasize that reusing the same password across multiple accounts can make it easier for hackers to gain access to multiple systems.

- **Log Out When Not in Use**: If your team works on shared computers or uses cloud services, encourage them to log out when they're finished, especially if they're working in a public space. It's a simple habit, but it adds an extra layer of security.

- **Install Software Updates Promptly**: Many people tend to ignore or postpone software updates, thinking they're just an inconvenience. But these updates often contain essential security patches. Remind your team that keeping software updated helps protect the whole business.

By creating these basic rules, you're establishing a foundation for safe online behavior that everyone can stick to without needing extensive training. The goal isn't to scare people but to encourage good habits that will protect both them and the business.

Lead by Example

If you want to create a culture of cyber awareness, it starts at the top. As a business owner or leader, your actions set the tone for the entire company. If you're following safe cybersecurity practices and talking about them openly, your team is much more likely to take them seriously.

For example, if you mention in a team meeting, "I changed my email password today to something stronger," it sends a subtle but powerful message: cybersecurity is important enough for everyone to prioritize, even the boss. You can also share stories about cybersecurity trends or incidents in the news to reinforce the importance of staying vigilant. By modeling good security behavior, you're showing that cybersecurity isn't just an afterthought—it's part of running a responsible and successful business.

Making Cyber Awareness Fun and Engaging

Let's be real—cybersecurity can seem dry or intimidating to many employees. But it doesn't have to be. Consider ways to make cybersecurity awareness engaging and even fun. Think of it like adding a little bit of gamification to your business's security efforts.

Here are a few ideas to bring some energy to cyber awareness:

- **Phishing Simulations**: Run occasional phishing simulations where you send out fake "phishing" emails to test how employees respond. No need to shame anyone who clicks the link—use it as a teaching moment instead. These simulations are a great way for employees to get hands-on practice in recognizing phishing attempts.

- **Friendly Cybersecurity Challenges**: Host a small competition where team members earn points for following good cybersecurity practices. For example, reward points for completing a cybersecurity training module, changing their password to something stronger, or spotting a phishing attempt in a simulation. Offer small rewards for those who participate, like a coffee gift card or an extra break.

- **Monthly Cybersecurity Tip of the Month**: Send out a "Cyber Tip of the Month" email to your team, where you share a short, practical tip related to security. It could be something as simple as, "Did you know? Using a phrase as your password can be stronger and easier to remember than a random jumble of letters."

When cybersecurity becomes part of your company's culture in a fun and engaging way, it feels less like a chore and more like a shared responsibility that everyone can contribute to.

Recognizing and Rewarding Good Cyber Habits

A little recognition goes a long way. When team members follow good cybersecurity practices, acknowledge it. It could be something as simple as saying, "Thanks for double-checking that email before clicking on the link. That's exactly the kind of vigilance we need to keep our data safe." By acknowledging and rewarding good behavior, you're reinforcing that cybersecurity is a priority.

Consider adding a "Cyber Champion" recognition once a quarter. Recognize the employee who's demonstrated good cybersecurity practices, whether they spotted a phishing attempt, kept their systems updated, or helped remind a colleague about a security rule. This can help motivate employees to stay aware and foster a sense of shared responsibility.

Building a Judgment-Free Reporting Culture

One of the biggest barriers to cybersecurity awareness is fear—fear of making a mistake, fear of admitting they clicked on something they shouldn't have, or fear of getting in trouble for a minor slip-up. To build a culture of cyber awareness, it's essential to create a judgment-free environment where employees feel comfortable reporting suspicious activity or admitting mistakes.

Encourage your team to come forward if they think they've encountered a phishing attempt or clicked on something suspicious. Emphasize that cybersecurity is everyone's responsibility, and mistakes can happen. The faster you know about potential threats, the quicker you can address them. In a judgment-free culture, employees are more likely to report incidents, allowing you to catch issues early before they escalate.

The Bottom Line: Cyber Awareness as a Collective Shield

Creating a culture of cyber awareness is about more than just training sessions or rules; it's about fostering a sense of collective responsibility. When everyone in the company feels empowered to make security-conscious decisions, your business becomes much harder to target. It's like having a team of watchful eyes, all looking out for potential risks, both big and small.

Remember, cybersecurity doesn't have to be scary or overwhelming. By making it a part of your everyday culture, you're building a workplace where everyone plays a role in keeping the business safe.

It's a journey, but with a few small steps, you can turn cybersecurity into something that everyone takes pride in—a collective shield that protects what you've worked so hard to build.

Step 5: Consider Bringing in Professional Help

By now, you've made a lot of progress. You've identified your digital assets, spotted weak points, set up basic defenses, and started fostering a culture of cyber awareness. You're already well ahead of the average small business when it comes to cybersecurity. But here's the thing: even with all the knowledge and good habits in the world, cybersecurity is an ever-evolving field. Threats are getting more sophisticated, and new vulnerabilities emerge all the time. And as a small business owner or leader, you've got enough on your plate. This is where professional help can make all the difference.

Cybersecurity isn't a "set it and forget it" kind of deal. It requires continuous monitoring, updating, and adapting to stay ahead of potential threats. Bringing in a cybersecurity expert or managed service provider (MSP) can give your business the level of security it deserves—without requiring you to become an IT whiz yourself. Let's explore why professional help is worth considering and how it can save you time, stress, and potentially even your business.

Why Consider Professional Help?

Think of it this way: you wouldn't try to fix a leaky roof on your own if you had no experience in roofing. Sure, you might be able to slap on a temporary patch, but you'd probably feel more secure if an expert handled it. Cybersecurity is similar. While basic practices can go a long way, a professional has the training, experience, and tools to spot vulnerabilities and plug gaps that you might miss.

Cybersecurity professionals spend their days studying the latest threats and learning how to counteract them. They're up-to-date on everything from new phishing techniques to emerging malware strains and even cutting-edge defensive technology. By bringing in an expert, you're leveraging all of that knowledge and experience to protect your business. It's like hiring a bodyguard who's trained to anticipate and deflect attacks—you can breathe easier knowing someone skilled has got your back.

Types of Professional Help: Cybersecurity Experts and Managed Service Providers (MSPs)

When we talk about "bringing in professional help," there are two main options: hiring a **cybersecurity consultant** or working with a **managed service provider (MSP)**.

1. **Cybersecurity Consultants**: These are independent professionals who specialize in assessing and strengthening your security posture. They'll typically start with a thorough evaluation of your business's current defenses, pointing out vulnerabilities and offering tailored recommendations. They can be hired for a one-time assessment or brought back periodically for ongoing support.

2. **Managed Service Providers (MSPs)**: An MSP is a company that provides ongoing IT support, including cybersecurity. They monitor your systems continuously, handle updates, and respond to threats in real-time. Essentially, they act as an outsourced IT team, but at a fraction of the cost of

hiring full-time staff. For small businesses without dedicated IT departments, an MSP can be a great option because they offer round-the-clock security support.

Both options have their advantages. If you're looking for a thorough check-up and expert advice, a consultant might be the right choice. If you prefer continuous protection and a proactive approach to security, an MSP can provide that peace of mind.

What Professional Help Can Offer That DIY Can't

While there's a lot you can do on your own, there are certain aspects of cybersecurity that really benefit from professional expertise. Here are a few ways that hiring an expert can take your security to the next level:

- **Comprehensive Security Assessment**: Professionals don't just look at surface-level issues; they dive deep. A cybersecurity expert or MSP will conduct a comprehensive security assessment, which may include vulnerability scans, penetration testing, and a review of your company's security policies. This thorough approach helps uncover hidden risks that may not be obvious to the untrained eye.

- **Advanced Threat Detection and Response**: Most small businesses don't have the resources or technology to constantly monitor for threats. But MSPs and cybersecurity experts do. They can set up systems that detect suspicious activity in real-time and respond immediately to potential threats. It's like having a night watchman for your digital space, ensuring that any intruder is caught before they can cause harm.

- **Tailored Cybersecurity Strategy**: Every business is unique, and so are its cybersecurity needs. A professional will customize a security plan based on your specific business operations, budget, and risk factors. For example, a retail store handling credit card transactions might need different protections compared to a law office storing sensitive client information. A one-size-fits-all approach won't cut it here, but a tailored strategy will.

- **Access to Specialized Tools and Technology**: Cybersecurity experts use specialized software and tools that aren't typically accessible to the average user. From advanced firewall configurations to intrusion detection systems, these tools are designed to provide robust, enterprise-level protection. MSPs often have partnerships with major security vendors, allowing them to offer high-quality tools at a lower cost than if you were to buy them individually.

- **Incident Response and Recovery**: No matter how well-prepared you are, there's always a chance that an attack could succeed. If that happens, it's essential to have a plan for minimizing the damage. Cybersecurity professionals are trained in incident response, meaning they can act quickly to contain a breach, identify how it happened, and help your business recover as efficiently as possible.

By bringing in an expert, you're essentially adding another layer of security to your business—one that's proactive, responsive, and specifically designed to handle whatever comes your way.

When Budget is a Concern: Prioritizing Your Needs

Let's be honest—cybersecurity experts and MSPs come with a price tag. But consider it an investment in your business's future. A cyberattack can cost thousands (or even tens of thousands) of dollars in recovery, lost revenue, and reputational damage. By investing in professional help, you're reducing the risk of incurring those costs.

If budget is tight, consider starting with a one-time security assessment from a consultant. This can give you a comprehensive view of your current security posture and a list of prioritized actions you can take. You can implement some of these actions on your own, then revisit the consultant for follow-up assessments as budget allows.

Alternatively, some MSPs offer tiered pricing plans. You might be able to start with basic monitoring and incident response, then add additional services as your business grows. Remember, even a little professional support can go a long way in strengthening your cybersecurity.

Finding the Right Fit: Questions to Ask Potential Experts

When you're ready to bring in professional help, choosing the right provider is essential. You want someone who understands the unique challenges of small businesses and can work within your budget. Here are a few questions to ask when interviewing cybersecurity consultants or MSPs:

1. **What experience do you have working with small businesses?**

    ○ Some cybersecurity experts specialize in large enterprises, while others are well-versed in the needs of small businesses. Look for someone who understands the specific challenges and constraints you face.

2. **What services do you offer, and how are they priced?**

   ○ Be clear on what services are included in their fee. Some MSPs offer "all-inclusive" packages, while others charge extra for certain services.

3. **How do you stay updated on the latest threats and trends?**

   ○ Cybersecurity is a fast-evolving field, so you want someone who actively stays current. Look for an expert who participates in ongoing training and certifications.

4. **Can you provide references or case studies?**

   ○ A reputable cybersecurity expert should be able to show you examples of how they've helped other businesses in situations similar to yours.

5. **What is your response plan in case of an incident?**

   ○ Ask how they handle emergencies. Knowing they have a clear incident response plan in place can give you peace of mind.

Making the Decision to Bring in Professional Help

Deciding to hire a cybersecurity expert or MSP isn't about admitting defeat—it's about recognizing that your expertise lies in running your business, not in defending it against cyber threats. And that's okay. Just like you wouldn't try to do your own dental work or legal filings, there's no reason to struggle through complex cybersecurity issues on your own when skilled professionals are available to help.

Bringing in professional help is about working smarter, not harder. With the right cybersecurity partner by your side, you're free to focus on what you do best: serving your customers, growing your business, and pursuing your passions. Professional help is there to fill in the gaps, providing the protection you need without diverting you from what you love.

Cybersecurity doesn't have to be an added burden. With the right support, it becomes another part of your business strategy—a proactive investment that keeps your business safe, your customers secure, and your peace of mind intact. Remember, the goal isn't just to survive in today's digital world but to thrive. And sometimes, having the right ally by your side makes all the difference.

Wrap-Up: Why Cybersecurity is a Continuous Process

Cybersecurity is Not "Set It and Forget It"

As you reach the end of this book, it's essential to remember that cybersecurity isn't something you can check off a to-do list and call it done. It's a continuous process—an ongoing commitment that, much like the daily routines that keep your business running, needs regular attention and care. Think of cybersecurity as a part of your business's "hygiene," a series of habits that keep your company clean and healthy in a digital sense.

When it comes to cybersecurity, there is no "set it and forget it." There's no magical button you can press that will safeguard your business against every possible threat forever. The digital landscape changes fast, and so do the methods and tactics of cybercriminals. Just like you wouldn't lock your physical doors once and assume your business is protected forever, you can't install a few digital locks and ignore them. You have to keep locking up every day, keep checking the windows, and keep an eye out for new vulnerabilities that may pop up.

Let's break down why cybersecurity is a never-ending journey and how keeping up with it can save you from countless potential headaches.

Cybersecurity: The Nightly "Locking the Doors" Routine

Imagine your cybersecurity routine as a nightly ritual, much like locking up the doors to your store before you go home. You wouldn't leave the doors open because "you locked them last week" or "you trust your neighborhood." Every night, without fail, you double-check the locks, make sure the windows are secure, and set the alarm if you have one. You do this because you know the cost of not securing your business each night is far too high.

The same principle applies to cybersecurity. Every day, your business interacts with a dynamic digital world full of risks and rewards. From emails to cloud storage, from your website to customer data, there are countless entry points that hackers might exploit. Just because you installed antivirus software last year doesn't mean you're safe today. Just because you updated your passwords six months ago doesn't mean they're impenetrable now. Cybersecurity requires regular "locking up" to ensure your business stays protected.

Why Cybersecurity Needs Regular Maintenance

Think of your cybersecurity setup like a car. You wouldn't expect a car to run perfectly if you never change the oil, rotate the tires, or take it in for maintenance. Over time, things wear out, parts need replacing, and new issues arise. Cybersecurity is similar. Technology changes, software gets outdated, and new vulnerabilities are discovered. A system that was secure last year might be riddled with holes today if it hasn't been maintained.

Here's the reality: cyber threats evolve. Hackers are constantly developing new techniques to bypass the very defenses that were effective last month. Just as you adapt your business strategies to changing markets and customer needs, your cybersecurity strategies must also evolve to keep up with new threats. Regular maintenance is not just a nice-to-have; it's a must-have if you want to avoid becoming an easy target.

Updating your systems, renewing software licenses, and reviewing access controls are all part of this ongoing maintenance. These tasks may seem tedious, but they're necessary. Much like you would inspect your physical space for hazards, cybersecurity upkeep helps you catch and resolve vulnerabilities before they turn into full-blown crises.

Building Cybersecurity Habits into Your Routine

To make cybersecurity a continuous process, it helps to build habits. Small, regular actions are far more effective than sporadic, intense bursts of effort. Cybersecurity can feel overwhelming if you approach it as a huge task that needs to be done all at once. But if you incorporate small actions into your regular routine, it becomes more manageable.

Consider setting a weekly or monthly schedule for cybersecurity tasks. Maybe every Monday, you review recent software updates, check for any unusual account activity, or remind your team to change their passwords if it's been a while. Perhaps every few months, you conduct a quick internal audit to ensure that access permissions are up to date and that former employees or vendors no longer have access to sensitive information. These small, consistent actions keep your cybersecurity sharp without disrupting your workflow.

In the same way that cleaning up clutter regularly prevents a big mess from building up, keeping up with cybersecurity tasks prevents small issues from turning into serious security breaches. By developing these cybersecurity habits, you're not just protecting your business today—you're creating a safer, more resilient business for the future.

Embracing the Mindset of Constant Vigilance

Effective cybersecurity requires a mindset shift. It's about recognizing that cybersecurity is not just an obligation; it's a vital part of running a modern business. Yes, the constant vigilance can feel like a burden, but think of it as a form of insurance. Just as you'd lock up your storefront to protect your inventory and your investment, ongoing cybersecurity vigilance protects everything you've worked so hard to build in the digital realm.

This vigilance isn't about living in fear of cyber threats; it's about being prepared and staying proactive. When cybersecurity is a regular part of your routine, it becomes second nature. You don't have to think twice about it, just like you don't think twice about locking the doors every night. The key is to stay aware of changes in the digital landscape, keep an eye on potential vulnerabilities, and be ready to adapt as needed.

And remember, you're not alone in this journey. Cybersecurity is a team effort. Encourage your employees to adopt the same mindset of vigilance. When your whole team is alert and aware, it creates a stronger, more resilient line of defense.

The Bottom Line: Cybersecurity is a Continuous Journey

As a small business owner, you wear many hats, and we know cybersecurity may feel like one more thing on an already full plate. But the reality is that cybersecurity isn't a one-and-done task—it's a continuous process. By approaching it as an ongoing commitment, you're protecting not just your data but also your reputation, your customer trust, and your business's future.

Yes, it takes effort. Yes, it requires time. But with each small step, you're building a digital fortress around your business, one that keeps evolving and strengthening against new threats. So, embrace cybersecurity as part of your business's rhythm. Make it a habit, stay vigilant, and keep locking those digital doors every day.

In the Wild West of today's digital landscape, the businesses that survive and thrive are the ones that stay alert, adapt to new challenges, and recognize that cybersecurity is a journey, not a destination. You're not just securing your present—you're paving the way for a safer, more successful future.

The Importance of Staying Informed

Cybersecurity is a fast-paced world, with new threats and technologies emerging all the time. As a small business owner, it might feel like keeping up with these changes is overwhelming, especially when your day is already packed. But here's the thing—staying aware of the latest cybersecurity news and trends doesn't have to be an all-consuming task. Even small steps, like skimming through a weekly newsletter or reading a quick article on recent threats, can make a big difference.

Why does staying informed matter? Because knowledge is power. Understanding what's happening in the cybersecurity world helps you recognize potential risks and anticipate new threats. It's like getting weather updates before a storm—knowing what might be coming allows you to prepare, even if it's just in small ways.

So, start simple. Subscribe to a few cybersecurity newsletters, follow reputable sources on social media, or set aside a few minutes each week to read up on the latest cyber news. Not only does this keep you aware, but it also empowers you to make smarter decisions for your business. By keeping cybersecurity top of mind, even in small doses, you're creating a proactive approach that can protect what you've worked so hard to build.

A Positive Note on Empowerment

As we wrap up this chapter, take a moment to appreciate how far you've come. Yes, the threats in the digital world are real, and yes, cybersecurity can feel daunting. But here's the powerful truth: every step you take, no matter how small, makes a difference. Every strong password you set, every backup you run, every time you pause before clicking an unfamiliar link—all of it adds up to a stronger, more resilient business.

Cybersecurity isn't about becoming invincible; it's about making yourself a harder target. And the beauty of it is, you don't have to tackle everything at once. Start with the basics, build good habits, and know that every small action contributes to a larger shield around your business.

Remember, you're not alone in this journey. Countless small business owners like you are facing the same challenges, and with each proactive choice, you're not just protecting your business—you're joining a growing community of empowered, cyber-aware leaders. You have the tools, you have the knowledge, and now, you have the mindset to keep your business safe in the digital age. Keep taking those steps forward, and know that each one brings you closer to a more secure future.

Understanding Your Digital Assets

What Are You Protecting?

Introduction to Digital Assets

When we talk about securing your business, what we're really discussing is safeguarding what matters most—your digital assets. But before you roll your eyes and think, *"Another tech term,"* let's break it down. A digital asset is, quite simply, anything of value to your business that exists in digital form. It could be customer data, financial records, product designs, intellectual property, or even that catchy tagline you came up with after three cups of coffee. If it's sitting in a file, a database, an email, or floating somewhere in the mystical realm of "the cloud," congratulations—it's a digital asset.

Now, identifying your digital assets is the first step in any cybersecurity journey, and here's why: if you don't know what you're protecting, how can you possibly protect it? Imagine you're a pirate—stay with me here—with a treasure chest full of gold and jewels. You wouldn't just toss it into a random room, lock the door with a rickety latch, and hope for the best, would you? Yet that's exactly what happens in the digital world far too often. Businesses scatter their most valuable information across hard drives, shared folders, and cloud servers without really knowing what's where, who has access, or how secure it is. It's digital chaos, and chaos is what hackers thrive on.

The stakes? Oh, they're high. Losing sensitive customer data is like a chef misplacing the secret sauce—it's bad for business and even worse for trust. Have your intellectual property stolen? That clever product design you spent months perfecting could suddenly be showing up in the hands of your competitors. Can't access financial records when you need them? You might as well run your business with a blindfold on. In short, overlooking your digital assets is a bit like leaving the treasure chest wide open with a neon sign that says, *"Help yourself!"*

But don't panic. This chapter is here to guide you through the process of identifying and organizing your digital treasure trove. Think of it as a modern-day treasure map, but instead of X's marking the spot, we'll show you how to pinpoint what's truly valuable, assess what's at risk, and keep track of everything without feeling like you're drowning in a sea of spreadsheets.

By the time you're done, you'll have a clear understanding of what digital assets matter most to your business, and you'll be well on your way to guarding them like the savvy treasure protector you are. So buckle up—it's time to tame the chaos and start building a cybersecurity strategy that works for you. After all, a treasure chest is only worth as much as the vault keeping it safe.

Mapping Out Your Assets

Before we dive into the thrilling world of digital locks and metaphorical security cameras, we first need to address a rather important question: *What, exactly, are you protecting?* Think of this as the inventory stage—the part where you open the digital vault, peer inside, and make a list of what's worth safeguarding. It's a bit like moving house, except instead of labeling boxes *"fragile"* or *"kitchen stuff,"* you're figuring out which pieces of data are vital to keeping your business running and which ones would make you break out in a cold sweat if they fell into the wrong hands.

Every business has assets worth protecting, even if you're not entirely sure what they are just yet. And here's the kicker: if you don't take the time to figure it out, you might as well hang a sign on your server that says, *"No need to knock—door's already open!"* Cybercriminals are more than happy to help themselves to anything left unguarded, and trust me, they're not picky about whether you know it's valuable or not.

So, let's roll up our sleeves and get to the heart of the matter: identifying what you're working so hard to build and protect. It's time to take stock, create a map of your digital treasures, and start thinking like a savvy business owner who understands that knowing your assets is the first step to keeping them safe.

Customer Data: The Crown Jewel

Ah, customer data. It's the lifeblood of your business, the digital equivalent of a crown jewel nestled in a velvet-lined box. And it's easy to see why—it's a treasure trove of names, addresses, phone numbers, credit card details, purchase histories, and perhaps even a few preferences (like whether Susan from Idaho prefers her tea Earl Grey or English Breakfast). These aren't just random bits of information;

they're the building blocks of trust, loyalty, and those all-important repeat purchases.

Now, here's the thing about jewels, digital or otherwise: when people entrust them to you, they expect you to guard them like a hawk. Nobody hands over their personal information with a shrug and thinks, *"Well, if it ends up in the wrong hands, that's just life."* No, they're counting on you to treat their data as if it's your own—maybe even better than your own, if you've ever misplaced your car keys for the third time in a week.

Cybercriminals, of course, are no strangers to the value of this digital bounty. To them, customer data is like a glittering heist waiting to happen. They'll swoop in, grab whatever they can, and vanish into the ether, leaving you to deal with the fallout—angry customers, potential lawsuits, and more sleepless nights than you care to count.

The takeaway here? Safeguarding customer data isn't just a good business practice; it's a solemn responsibility. In the digital realm, these bits and bytes are as valuable as diamonds and pearls, and they deserve the same level of care. So, lock it up, protect it fiercely, and remember: in this game, you're not just running a business—you're guarding the crown jewels of trust and customer loyalty.

Financial Records: Protecting the Heartbeat of Your Business

If customer data is the crown jewel of your operation, then financial records are its steady, dependable heartbeat. This is the stuff that keeps your business alive: invoices, payroll, tax documents, cash flow records—essentially, the digital equivalent of your company's pulse. Without them, the whole operation grinds to an unceremonious halt, and nobody wants to see that.

To a hacker, though, financial records are like finding a golden ticket in a candy bar. This isn't just data; it's high-value, high-demand treasure with an eager market of buyers. These aren't just numbers on a spreadsheet, oh no. They're the lifeblood of your company, the threads holding the whole operation together. Mess with those, and suddenly the gears stop turning, and you're left staring at a very bleak bottom line.

And yet, financial records often get the short end of the stick when it comes to security. They're not glamorous, like a sparkling new product launch or a clever marketing campaign. But they're absolutely essential. Without them, there are no launches, no campaigns—just chaos. So, give them the care they deserve. Lock them down, back them up, and guard them as fiercely as you would your business's actual heartbeat. Because in many ways, that's exactly what they are.

Employee Information: Don't Overlook Your Team's Privacy

Your employees aren't just people who show up, punch a clock, and call it a day—they're the heartbeat of your business. They make everything run smoothly, handle the challenges, and occasionally bring in cake on Fridays (a highly underrated contribution). So, just as you'd go to great lengths to protect your business, you owe it to your team to protect their personal information too.

Employee information might include names, addresses, salary details, and even health records. It's the kind of sensitive data that deserves the same level of vigilance as customer information, if not more. Because here's the thing: if this data gets exposed, you're not just looking at a potential legal mess. You're dealing with a trust issue, and rebuilding that with your team is no easy task. Safeguarding their information shows your employees that their privacy matters and that you value their well-being beyond the hours they spend on the job.

Intellectual Property: Guarding What Makes You Unique

Now, let's move on to the fun stuff: intellectual property. This is the crown jewel of creativity—the secret sauce that makes your business, well, yours. It includes your unique processes, product designs, trade secrets, and all the brilliant ideas you've poured countless hours into.

Picture this: your competitors suddenly get their hands on your best ideas. It's a nightmare scenario. Protecting your intellectual property isn't just about safeguarding your competitive edge; it's about protecting the soul of your business. These are the things that make you distinct, the culmination of your ingenuity and hard work. Losing them isn't just a financial blow—it's a loss of identity. So, lock it down, guard it fiercely, and keep it close. Because once it's out there, there's no getting it back.

Business-Critical Document: The Unassuming Backbone

Finally, let's talk about the unsung heroes of your business—those other essential documents that keep the wheels turning. Contracts, legal agreements, supplier paperwork—they might not have the dazzle of intellectual property or the sensitivity of employee data, but they're every bit as important.

Without these documents, running your business would be like trying to assemble IKEA furniture without the instructions—possible, but wildly inefficient and likely to end in frustration. These might not scream *"Top Secret!"* but they're the foundation that holds your operations together. Protecting them isn't about locking them away in a high-tech vault; it's about ensuring they're safe, accessible, and secure, so your business can keep humming along without a hitch.

In the end, it's about treating every piece of your business—people, ideas, and processes—with the care and respect it deserves. Because in the grand scheme of things, they're all part of what makes your operation not just functional, but exceptional.

The Importance of Documenting Your Assets

Let me ask you this: if someone were to take a random inventory of your business right now—every digital asset, document, file, and idea—how confident would you be that you could account for all of it? If your answer is somewhere between *"Reasonably confident"* and *"Oh dear, not very,"* then you're not alone. For many businesses, the idea of documenting their assets feels about as appealing as organizing a sock drawer—it's tedious, not particularly exciting, and you'd rather be doing almost anything else.

But here's the thing: if you don't know what you have, how will you ever know what you've lost? Documenting your assets is like creating a map of your digital treasure. Without it, you're wandering blindfolded through a maze, hoping to stumble upon what matters. And in a world where cyber threats are lurking around every corner, that's not exactly the safest strategy.

So, before we dive into the nuts and bolts of asset documentation, let's pause to appreciate why this step matters so much. It's not just about satisfying your inner neat freak (though that's a bonus); it's about ensuring your business is prepared, protected, and equipped to handle whatever comes its way. After all, you can't protect what you don't know exists—and when it comes to your digital assets, ignorance isn't bliss. It's a liability.

Risk Assessment: Asking the Big "What Ifs"

"Risk assessment" might sound like one of those imposing phrases tossed around in corporate boardrooms by people in sharp suits with PowerPoint slides. But at its core, it's really just a grown-up version of asking, *"What if...?"* And who doesn't enjoy a good round of *"What if...?"* It's your chance to let your imagination run wild, in a constructive way, of course, and ask yourself a few straightforward, albeit slightly unsettling, questions:

- *What if this data gets leaked?* How would it affect my business, my customers, or my sanity?

- *What if it's accidentally deleted?* Do I have a way to recover it quickly, or am I about to embark on a long and harrowing journey into the depths of regret?

- *What if it's stolen or altered?* What sort of chaos would that unleash?

Here's the good news: you don't need to be a technical wizard or have an IT degree to figure this out. A risk assessment isn't about diving into the deep end of cybersecurity jargon—it's about applying some good old-fashioned common sense. Think of it like preparing your house for a big storm. Are the windows locked? Is there a backup generator in case the power goes out? Have you tied down anything that might blow away?

In the same way, a *"what-if"* analysis for your cybersecurity helps you spot the weak points and figure out how to shore them up before the storm hits. It's not about creating a fortress—it's about making sure you're ready for whatever comes your way. And let's be honest: peace of mind is worth a little bit of planning, don't you think?

Assigning Value to Each Asset

Now let's talk about value—because, as much as we'd like to think everything in our digital treasure chest is equally precious, the truth is, it's not. Some assets are the crown jewels, while others are more like the spare change under the sofa cushions. Understanding the value of each is crucial for deciding how much time, effort, and resources you should devote to protecting them.

Assigning value doesn't have to involve spreadsheets or algorithms (though if you enjoy those things, by all means, have at it). Instead, it's a simple matter of asking yourself two key questions:

- **What would this cost me in dollars and cents?** Imagine a data breach. How much would it set you back in fines, lost revenue, and the headache-inducing expenses of recovery? Think about it as the equivalent of your plumbing bursting—you're not just paying to fix the pipes; you're replacing the carpet and possibly rethinking your life choices.

- **What would this cost me in trust and reputation?** Here's where it gets trickier. Some assets—like customer data or trade secrets—don't just carry a financial price tag. They hold intangible value. If customer data leaks, the financial impact might be measurable, but the loss of trust? That's the gift that keeps on giving (in the worst possible way). Reputation, as they say, takes years to build and seconds to ruin.

In short, assign a dollar amount where you can, but don't overlook the priceless stuff—the goodwill of your customers, the unique edge that sets your business apart, and the sense of security you've worked so hard to establish. Losing these can hurt just as much, if not more, than a financial blow. And let's face it, they're a lot harder to get back once they're gone.

Evaluating the Impact of Loss

Let's put this into perspective with a little hypothetical scenario. Imagine, if you will, that a data breach has occurred, and your customer information—names, email addresses, and, heaven help us, credit card details—is suddenly out in the wild. In the immediate aftermath, chaos ensues. You're scrambling to figure out what went wrong, facing potential fines, sending apologetic emails to customers, and likely calling in a cybersecurity consultant who charges by the hour (and not a small hourly rate, at that).

But here's the rub: that's just the beginning. Customers, understandably miffed, might start questioning your commitment to keeping their information safe. Some may decide to take their business elsewhere, muttering unkind things about you as they go. And if the breach makes its way into the public domain—because let's face it, these things always do—your reputation could take a nosedive. Suddenly, this one breach has morphed into a hydra-headed problem, threatening not just your current customer base but also future prospects.

Now, this exercise isn't meant to send you into a spiral of existential dread. It's not about being pessimistic; it's about being prepared. By evaluating the potential impact of loss, you gain a clear-eyed understanding of what's at stake, which is essential for making the case that cybersecurity isn't just a line item on your budget—it's a priority.

Prioritizing Based on Vulnerability and Value

Once you've taken stock of what's valuable and what's vulnerable, it's time to do what every sensible person does in a crisis: prioritize. Not all assets are created equal, and some are far more at risk than others. These high-value, high-vulnerability items should shoot straight to the top of your protection list. Others, while important, may be less critical or harder for hackers to reach, earning them a spot lower down the list—but don't neglect them entirely.

Think of this process as digital triage. Just like in an emergency room, you start with the most critical issues. For example, if customer data is both incredibly valuable and sitting in an easily accessible, poorly protected location (perhaps guarded only by the cybersecurity equivalent of a wet paper bag), that's an obvious candidate for immediate action.

Prioritizing in this way isn't just about throwing money or resources at the problem. It's about being strategic. You're focusing your efforts on what matters most, shoring up the areas where a breach would have the biggest impact. This isn't just reactionary fire-fighting; it's proactive, thoughtful defense-building that fortifies your business for the long haul. And, let's be honest, wouldn't it feel nice to sleep at night knowing the digital shop is secure?

Creating a "Cyber Safe" List: Prioritizing What Matters Most

Now that you've bravely mapped out your digital treasure trove and identified the weak spots, it's time to get strategic. Enter the "Cyber Safe" list—your very own emergency action plan for the digital world. Think of it as a fire drill for your data, but without the awkward sprint to the parking lot. The goal here isn't to protect *everything* equally (as noble as that sounds). Instead, it's about focusing your resources on the assets that matter most, the ones that would cause the most chaos if they fell into the wrong hands.

This isn't as daunting as it might sound. In fact, it's wonderfully straightforward. The idea is to rank your digital assets by importance and vulnerability, so you know exactly where to direct your efforts. By doing this, you'll ensure that your most critical data—the crown jewels of your digital domain—are given the care and attention they deserve.

Ready to dive in? Let's roll up our sleeves and build your "Cyber Safe" list, one asset at a time. Because in the world of cybersecurity, prioritization isn't just practical—it's survival.

What is a "Cyber Safe" List?

Picture your Cyber Safe list as the VIP section of your digital domain—a prioritized roster of your most important assets that deserve the best security money (and a bit of common sense) can provide. Much like a fire evacuation plan for a building, this list is your go-to guide when things heat up in the cyber world. The idea is simple: when disaster strikes, you know exactly where to start, ensuring your most valuable data isn't left flailing in the chaos.

The goal here is as straightforward as it is vital: identify the digital lifeblood of your business—the assets you simply cannot live without—and give them the top-tier treatment. I'm not talking about locking everything away in Fort Knox here. What we're really heading toward is putting your critical data under the equivalent of a digital spotlight, so you know where to focus your efforts when it matters most.

Identifying the Most Critical Assets

First, let's start with some honesty: what are the assets your business couldn't function without? This isn't the time for modesty. Customer data? A no-brainer. Financial records? Absolutely. Proprietary designs, trade secrets, or that genius marketing idea you sketched on a napkin? Without a doubt.

These high-value assets should sit proudly at the top of your Cyber Safe list and receive the gold-standard of protection. Think encryption, secure backups, multi-factor authentication, and anything else that makes hackers mutter under their breath and move on to easier prey. Not everything in your digital universe needs this level of care, but the items that make or break your business? They certainly do.

Organizing Assets by Sensitivity

Once you've singled out your VIPs, it's time to get organized. Not all data is created equal, so let's categorize it into tiers like "High," "Moderate," and "Low" sensitivity. Think of it as sorting your wardrobe—your most prized possessions (that designer jacket, metaphorically speaking) go in the safest spot, while the old sneakers can sit in the corner.

- **High Sensitivity**: These are the heavy hitters—the data that, if compromised, could cause catastrophic financial, reputational, or legal damage. Customer credit card details, confidential contracts, regulatory documents—they all go here. This is where you pull out the cybersecurity big guns.

- **Moderate Sensitivity**: Important but not quite DE-FCON-1 level. These might include internal reports, day-to-day communications, or marketing materials. Losing them would be inconvenient, but it wouldn't send your business into a tailspin.

- **Low Sensitivity**: These are the digital equivalents of public-facing shop signs—nice to have secure but unlikely to cause a meltdown if compromised. General announcements or publicly available documents often land in this category.

By sorting your assets this way, you create a clear roadmap for where to direct your resources. High-sensitivity items get the lion's share of attention, while moderate and low-sensitivity data can be handled with a more balanced approach. It's practical, effective, and, dare I say, oddly satisfying.

In the end, a well-organized Cyber Safe list is about peace of mind as much as it is about protection. And really, isn't that what we're all after?

Where to Store Critical Assets Safely

Now that you've identified the most precious jewels in your digital crown, the next logical step is deciding where to stash them. And no, stuffing them in a metaphorical sock drawer isn't going to cut it. Not all storage solutions are created equal, and the higher the stakes, the more thought you'll want to put into their safekeeping. Let's explore your options:

- **Encrypted Cloud Services**: Think of this as the high-tech equivalent of a safety deposit box. For assets that need to be accessible yet highly secure, encrypted cloud storage is a fantastic choice. It's like adding a secret code to your valuables—if someone manages to access your account, they'd still need the encryption key to make sense of anything. It's convenient without sacrificing security.

- **Secure Offline Backups with Encryption**: For your most sensitive data—the kind of information that could sink your business if compromised—a highly secure offline backup is a must. This doesn't mean just tossing files onto an external hard drive. Use an encrypted external storage device specifically designed for sensitive data, ensuring that even if the device is physically stolen, the information remains inaccessible without the encryption key. Store this device in a secure, locked location, such as a fireproof safe, and update the backup regularly. Offline backups should also follow the 3-2-1 rule: three copies of your data, stored on two different mediums, with one copy stored securely offsite. This approach ensures both safety and redundancy, protecting you from threats like ransomware and hardware failure.

- **Multi-Layered Security Systems**: For your top-tier assets, consider storing them in systems with multiple layers of security. We're talking multi-factor authentication, biometric access, and protocols that would make a spy thriller look dull. The more valuable the data, the more hoops hackers should have to jump through to get anywhere near it.

It's really just like deciding where to stash your most valuable jewelry at home. Some pieces can live in a regular jewelry box, but the crown jewels? They belong in a safe. Your digital assets deserve the same level of respect and care.

Establishing Access Control for Each Asset

Now, let's talk about who gets the keys to the digital vault. Just as you wouldn't hand out spare house keys to every passing acquaintance, you shouldn't dole out access to your digital assets willy-nilly. This is where access control swoops in to save the day.

For each item on your Cyber Safe list, take a moment to think: *Who really needs access to this?* Your customer service team might need customer records to do their job, but do they need payroll data? Probably not. By limiting access to only those whose roles genuinely require it, you significantly reduce the chances of accidental mishaps—or worse, intentional breaches.

Access control goes further than just locking people out; in order to do this correctly, you have to have a clear, and well documented plan for setting boundaries and minimizing unnecessary risks. It also adds a handy layer of accountability. When only a select few have access, it's much easier to track who's doing what with your data. Think of it as drawing a neat, sensible fence around each asset—keeping things secure while ensuring everyone stays in their lane.

Wrapping It Up: Your Cyber Safe List in Action

By building a Cyber Safe list, categorizing your assets by sensitivity, choosing secure storage solutions, and establishing clear access controls, you're not just checking off a cybersecurity to-do list—you're creating a thoughtful, strategic defense plan.

This is about more than just protecting data. It's about safe-guarding the very foundation of your business. By taking these steps, you're ensuring that when the digital winds blow—and they always do—your most valuable assets remain secure, accessible, and exactly where they should be. One asset at a time, you're fortifying your business against whatever challenges come its way.

Tracking Your Digital Assets: Staying Organized Without the Overwhelm

Congratulations—you've crafted your Cyber Safe list and identified your digital treasures. But here's the thing: knowing what you have is only the first step. Keeping track of those assets is just as important, and let's be honest, it's easy to let things slip into chaos. That said, tracking your assets doesn't have to feel like tackling a pile of mismatched socks. With a bit of structure and a few handy tools, you can stay on top of your digital inventory without drowning in details.

Why Asset Tracking is Important

Let's face it: the phrase *"tracking assets"* sounds about as thrilling as alphabetizing your spice rack. But here's the good news: you don't need to overdo it. Tracking is about getting a clear picture of what you have, what's critical, and where it all lives. After all, if you don't know what you're trying to protect, how can you protect it?

This isn't about cataloging every digital file since the dawn of time. Instead, it's about making sure the important stuff—customer data, financial records, intellectual property—is accounted for. Trust me, a little effort now will save you from a lot of frantic scrambling later when you're trying to locate that one crucial file in the middle of a cyber crisis.

DIY Inventory: Spreadsheets and Checklists

If you're running a small business on a budget, don't worry—you don't need fancy software to keep track of your assets. Sometimes, the simplest tools are the best. A basic spreadsheet or checklist can do the job beautifully.

Here's a structure to get you started:

It's simple, straightforward, and easy to update. If spreadsheets make you want to run for the hills, no worries—use a notebook, a checklist app, or even sticky notes. The goal is to create an overview of what's valuable and where it's located so you're never caught off guard.

Software Solutions for Asset Management

For those with a growing list of assets—or a deep love of automation—there are plenty of asset management tools to lighten the load. These programs do the heavy lifting, keeping track of assets, updating records, and even sending alerts when something changes.

User-friendly options like Asset Panda, ManageEngine, or Spiceworks are great for small businesses. They provide an easy way to stay organized without needing a PhD in IT. The trick is finding a tool that fits your needs and your budget. You don't need a Swiss Army knife when a sturdy screwdriver will do.

Best Practices for Keeping Inventory Up-to-Date

Now that your inventory is in place, don't let it gather digital dust. The cyber world changes fast—new assets appear, old ones become obsolete, and permissions shift. Set a calendar reminder to review your inventory every quarter. It doesn't have to be a monumental task; just a quick check-in to ensure everything is current.

Think of it like cleaning out your fridge. Every now and then, you need to toss out the expired yogurt and make room for the fresh stuff. A well-maintained inventory not only keeps you organized but also helps you spot security gaps before they become problems.

Tools for Organizing Access Controls

Finally, let's talk about who gets the keys to the digital castle. Managing permissions might sound like a bureaucratic nightmare, but modern tools can make it at least somewhat simple. Platforms like Google Workspace and Microsoft 365 let you control who can access, edit, or share specific files with just a few clicks.

These tools come with built-in permissions management, so you don't need to be a tech wizard to set up access controls. When an employee leaves or switches roles, you can update their access rights in moments. It's like playing digital gatekeeper, ensuring only the right people get into the right rooms.

With a solid tracking system, regular updates, and sensible access controls, you're not just staying organized—you're building a strong foundation for cybersecurity. It's not just about defending your data; it's about creating a system that works for you, keeps your assets safe, and lets you sleep soundly at night.

Wrapping It Up: A Little Organization Goes a Long Way

Let's be honest—tracking your digital assets probably doesn't sound like the most riveting task on your to-do list. But here's the thing: with a bit of structure, a smidgen of effort, and a sprinkle of automation if you're feeling fancy, you can turn what feels like a chore into a game-changer for your cybersecurity strategy. It's not about cataloging every file as though your business were a museum; it's about knowing what's valuable, where it lives, and who's allowed to touch it. With that knowledge, you're not just managing your assets—you're safeguarding them like a pro.

Putting It All Together: Your Asset Protection Checklist

Alright, you've come this far, so let's bring it all together with a handy-dandy checklist. Think of this as your cybersecurity starter kit—a simple, effective way to organize, secure, and manage your digital valuables. It's straightforward, no-nonsense, and designed to keep you sane while keeping your business safe.

Checklist Recap

Here's a quick rundown of the main steps we've covered in this chapter. Use this checklist to make sure you've got everything covered, one step at a time:

1. **Map Your Assets**: Identify all the digital assets your business relies on—from customer data and financial records to intellectual property and employee information.

2. **Assess What's at Risk**: Take stock of potential risks for each asset. Ask yourself, "What could go wrong?" and consider the impact of a data breach, loss, or unauthorized access.

3. **Prioritize by Value and Vulnerability**: Rank each asset based on how valuable it is to your business and how vulnerable it might be. Those high-value, high-risk assets are the ones to secure first.

4. **Track and Document**: Start a basic inventory of your assets, using a spreadsheet, checklist, or asset management software. Keep it simple but thorough, and don't forget to update it regularly.

5. **Create Your Cyber Safe List**: Mark down your most critical assets that need extra protection. Think of this as your "fire drill" list—the things you'd grab first in an emergency.

Keeping this checklist handy will ensure that you stay on track as you start to build a solid foundation for your digital asset security.

Sample Asset Inventory and Cyber Safe List

To give you an idea of what a finished product might look like, here's a sample of a basic asset inventory and Cyber Safe List. You can use this as a reference when creating your own, customizing it to fit the unique needs of your business.

Sample Asset Inventory:

Sample Cyber Safe List:

1. **Customer Database** – This is priority number one. It contains sensitive personal information and is crucial to keep secure.

2. **Financial Records** – The heartbeat of your business. Invoices, tax records, payroll—all things you can't afford to have compromised.

3. **Product Designs** – Your unique ideas and intellectual property deserve extra protection. These keep your competitive edge intact.

4. **Employee Records** – Protecting your team's privacy is just as important as protecting customer data.

This simple inventory helps you visualize where each asset stands in terms of value and risk. Feel free to tweak this template to fit your business. The key is to ensure that your highest-value assets are front and center on your Cyber Safe List, getting the protection they deserve.

Next Steps for Asset Protection

Now that you have the tools, it's time to roll up your sleeves and start mapping out your own digital assets. Don't worry if it's rough at first—the goal here is to get started. Begin with a basic list of your most critical assets, and as you go, add details and refine your inventory. This chapter is just the beginning. In future sections, we'll dive deeper into strategies for safeguarding your digital valuables, so you'll build on this foundation as you go.

Remember, protecting your business isn't about making everything perfect right off the bat. It's about taking small, manageable steps that, over time, create a strong shield around your assets. So grab that checklist, start your inventory, and know that each step you take is one step closer to a more secure, resilient business.

Conclusion: Understanding Your Assets is the Key to Protection

As we wrap up this chapter, let's take a moment to reflect on how far you've come. When we first started, you might have been wondering, *"What assets?"* and perhaps feeling a bit unsure about where to even begin. Now, look at you—you've created a roadmap of your business's digital valuables, clearly identifying what matters most. That's no small feat. In fact, it's a giant leap forward in the often murky world of cybersecurity.

But why does this really matter? Because understanding your assets is the cornerstone—the bedrock, even—of any effective cybersecurity plan. Protecting your business is a lot like guarding a treasure: you can't very well keep it safe if you don't know where it's buried. By identifying what's critical to your business—whether it's customer data, financial records, or the intellectual property that makes your company unique—you're equipping yourself to make smart, targeted decisions about where to focus your efforts.

And here's the beauty of it: this isn't just about ticking boxes or following some abstract advice. This is about taking real, practical steps to shield your business from threats. It's about putting yourself in control. So, take a deep breath, pat yourself on the back (seriously, you deserve it), and know that you're already miles ahead of where you started.

The Importance of Knowing Your Assets

Think of your digital assets as the valuables in a safe deposit box. If you don't know what's in there, how can you possibly protect it? The same applies to your digital world. Knowing what you have and where it lives is the foundation of any good cybersecurity strategy. It's hard to defend a treasure if you're not even sure what it looks like, let alone where it's stashed.

By clarifying what's valuable in your business—be it customer data, financial records, or those brilliant product designs you spent months perfecting—you're setting yourself up for success. Awareness isn't just helpful; it's powerful. It allows you to move from reacting to problems as they arise to proactively defending what's most important. In short, knowing your assets gives you the upper hand, and in the world of cybersecurity, that's no small advantage.

The Power of Prioritization

Here's the thing: not all assets are created equal. Just as some parts of your business demand more attention than others, some digital assets are far more critical. Maybe it's your customer database, the backbone of your sales operations. Or perhaps it's your financial records, the lifeline that keeps your company running. Whatever they are, these high-value, high-risk assets deserve your strongest defenses.

Prioritization isn't about ignoring the smaller stuff; it's about being strategic. It's about putting your resources where they'll have the biggest impact. By focusing your efforts on the assets that matter most, you're not just staying organized—you're staying confident. You're ensuring that your business is protected where it counts, without feeling like you need to lock down every last file with the intensity of Fort Knox. This isn't just practical; it's empowering.

Look Ahead to the Next Chapter

Now that you've got a solid understanding of what needs protecting, it's time to turn our attention to a slightly more sinister subject: cybercriminals. In the next chapter, we'll dive into the minds of these digital outlaws—what drives them, what they're after, and the tactics they use to get it.

It might sound daunting, but here's the good news: understanding your enemy is half the battle. By learning how they operate, you'll be even better equipped to shore up your defenses and protect the treasures you've worked so hard to build. And trust me, everything you've learned so far will feel even more vital once you see how these cyber miscreants ply their trade.

With this foundation in place, you're well on your way to becoming a more resilient, cyber-savvy business owner. So take another deep breath (it's good for you), give yourself another well-earned pat on the back, and get ready for the next chapter. You're doing fantastically, and the best is yet to come. Let's keep going—you've got this!

THE CYBER CRIMINALS' TOOLBOX

HOW HACKERS GET IN

Introduction to the Hacker's Toolbox

Welcome to the shadowy and slightly unnerving world of cy-bercriminals—a realm where cunning, creativity, and code collide. If you're going to mount a proper defense against the digital rogues of the modern age, you first need to understand who they are and how they operate. Think of this as getting a peek into their toolbox—a collection of sneaky tricks, sophisticated gadgets, and audacious strategies designed to exploit the unwary.

Who are these hackers, exactly? Well, they're a diverse bunch. Some are the quintessential shadowy figures, hunched over keyboards in dimly lit rooms, their screens glowing with mysterious lines of code. Others are part of organized groups, plotting their next big score like modern-day heist crews, only without the balaclavas and getaway cars.

And then there are the big players—nation-states wielding cyberattacks as weapons of influence, destabilization, or just plain old chaos.

Make no mistake: these aren't your garden-variety troublemakers poking around for fun. These are skilled operators—individuals and teams—dedicated to finding vulnerabilities and exploiting them, often in businesses that aren't quite as prepared as they should be. They're the kind of people who see your overlooked software update or weak password and think, *"Ah, delightful!"*

I'm not trying to scare you into sleepless nights, but I do want you to have a deep enough understanding about cybercriminals to arm yourself with knowledge. Because in the world of cybersecurity, knowing how the enemy operates is half the battle. So, buckle up, and let's dive into the minds of the digital outlaws lurking just beyond your firewall. It's a journey that will leave you informed, empowered, and better prepared to protect what matters most.

Understanding the Enemy: Who are Cybercriminals?

Let's dispel a popular myth right out of the gate: not all cybercriminals are loners hunched over laptops in dimly lit basements, fueled by energy drinks and mischief. The reality is both more sophisticated and, frankly, more unsettling. Cybercriminals are a varied bunch, with motives and methods as diverse as their hideouts (some of which, I'm sure, are better lit than we imagine).

First, you've got the solo operators. Think of these as the hobbyists of the hacking world. For some, it's a challenge—a digital puzzle to crack. For others, it's a thrill, a modern-day equivalent of picking locks just to prove they can. Then there are the organized crime rings, which operate with the coordination and efficiency of a well-oiled machine. These aren't ragtag teams; they're multinational operations with hierarchies, resources, and even business plans—albeit highly illegal ones. And let's not forget the nation-states, where cybercrime takes on a more menacing role. These are the players who use hacking as a tool of geopolitics, targeting infrastructure, corporations, and governments with attacks that are as strategic as they are destructive.

What ties all these groups together? A shared, unrelenting goal: to exploit weaknesses and gain unauthorized access to data, money, or resources. And here's the kicker—while it's easy to assume they're only interested in big targets like major corporations or state secrets, that's far from the truth. Cybercriminals love small businesses. Why? Because smaller operations often lack the robust defenses of their larger counterparts. To these hackers, small businesses are like houses with unlocked doors in an otherwise gated neighborhood—tempting, accessible, and easy to exploit.

So, while the imagery of lone hackers and shadowy figures has a certain cinematic charm, the reality is far more nuanced—and, for business owners, far more concerning. Knowing your enemy means understanding the variety of faces behind the screens, from the solo tinkerers to the well-funded syndicates. It's a landscape as complex as it is perilous, but by grasping the scope of who you're up against, you're already taking a critical step toward protecting your business.

Why Knowing Their Tools Helps You Defend Your Business

Here's the silver lining in the dark cloud of cybersecurity: you don't have to be a tech wizard to defend your business against hackers. But understanding the basics of how these digital rogues operate? That can make all the difference. Think of it like learning a pickpocket's tricks—once you know how they work, you're far less likely to have your wallet spirited away in a crowded marketplace. The same principle applies to cyber threats. By grasping the tools and tactics hackers rely on, you'll start spotting vulnerabilities before they morph into full-blown crises.

Throughout this chapter, we're going to take a peek into the hacker's toolbox. Don't worry, this isn't about inducing panic or turning you into a sleepless bundle of nerves. It's about empowering you. We'll explore the methods, software, and social engineering tricks that hackers use to worm their way into digital systems. Armed with this knowledge, you'll be better positioned to shore up your defenses and make your business a far less attractive target.

Think of it as leveling the playing field. Hackers count on ignorance—on businesses not knowing what's out there or how to protect themselves. But once you start to understand how these tools work, you shift the balance. Suddenly, you're not just reacting to threats; you're anticipating them, blocking them, and keeping your business's digital drawbridge firmly raised.

So buckle up—we're about to dive into the hacker's world, explore their tools, and demystify their strategies. The more you know about how hackers think, the harder you make it for them to get in. And that, as you'll soon see, is a victory worth aiming for.

Social Engineering: When Hackers Play Mind Games

What is Social Engineering?

Picture a hacker who doesn't bother with firewalls, antivirus software, or any of those pesky technical hurdles. Instead, they take the simpler, sneakier route: convincing someone to hand over the keys to the castle.

This is the essence of social engineering—a tactic as old as human nature itself. Rather than relying on coding wizardry or brute force, these cybercriminals use good old-fashioned psychological manipulation to access sensitive information. And the worst part? The people they manipulate often have no idea they've been duped.

Think of social engineering as the hacker's equivalent of a con artist's hustle. Why wrestle with complicated security systems when you can simply trick a well-meaning employee into opening the door for you? These criminals don't need to be technical geniuses; they just need to understand human behavior. And oh, how they do. They craft convincing backstories, impersonate trusted colleagues, or prey on emotions like urgency, curiosity, or helpfulness.

Imagine an email that looks like it's from your CEO, urgently asking for login credentials to "resolve a critical issue." Or a call from someone claiming to be IT support, requesting access to your system to "fix a problem." It's plausible, believable, and designed to disarm suspicion. Once trust is gained, the hacker has all they need to wreak havoc—no complicated hacking required.

Social engineering isn't just a digital trick; it's a psychological game. It's about exploiting the natural inclination to trust, to help, and to act quickly when something seems urgent. And the results can be devastating, leaving businesses wondering how their tight security measures were bypassed without a single line of code being cracked.

It's clever, it's devious, and unfortunately, it's highly effective. The key to combating it? Awareness. Because once you know the game, it's much harder to be played.

Why Social Engineering Works

Here's the thing about social engineering—it works because it taps into our most basic, endearing human traits. Most of us like to be helpful. We're curious by nature. And when someone shows up looking and sounding like they know what they're doing, we tend to trust them. Cybercriminals know this. In fact, they count on it. They don't need to be tech wizards or criminal masterminds; they just need to be convincing enough to turn those very human qualities into vulnerabilities.

Even the most cautious, well-meaning employees can fall for social engineering. And why wouldn't they? These attacks don't feel like attacks. The hacker doesn't show up twirling a virtual mustache and cackling maniacally. No, they come disguised as a friendly IT technician, a sympathetic colleague, or perhaps a worried customer in need of immediate assistance. It's the digital equivalent of a wolf in sheep's clothing, only the wolf also knows exactly how to sound polite, professional, and just a tad urgent.

The brilliance—and the danger—of social engineering lies in how disarmingly natural it feels. The hacker's request is crafted to seem reasonable, even routine. Resetting a password? Sure, that makes sense. Sending over a file? Of course, happy to help. And before you know it, the hacker has exactly what they wanted, leaving you none the wiser until the damage is done.

Social engineering doesn't rely on complex code or brute force. It relies on people being people—helpful, trusting, and just a bit too quick to act without questioning. And that's what makes it so effective. But fear not: once you learn to spot the signs, you'll see these attacks for what they really are—suspicious, manipulative, and absolutely preventable.

Common Social Engineering Tactics

So, how exactly do hackers pull off these mind games? Let's look at a few of the most common tactics they use:

- **Pretexting**: This is where the hacker creates an elaborate backstory, or "pretext," to justify their request for information. Imagine someone calls your office claiming to be from a third-party vendor, saying they urgently need your account details to "process a payment." Or, they pose as a high-ranking executive, insisting they need access to certain files for an "important meeting." The pretext gives the hacker a believable reason to ask for sensitive information, and they rely on the recipient's instinct to trust people in authority or those with plausible stories.

- **Impersonation**: Hackers are masters of disguise—digital ones, at least. In an impersonation scam, they pretend to be someone trusted within your business. Maybe they spoof an email to look like it's coming from the CEO, asking a junior employee to transfer funds to a new vendor. Or, they pose as an IT technician, calling to "fix an issue" with your computer, all the while guiding you into granting them access. This tactic preys on the assumption that if a message or call seems to come from a legitimate source, it must be safe.

- **Baiting**: Here, the hacker dangles something enticing—"bait"—to lure someone into a trap. It could be a free USB drive left on a desk labeled "Confidential" (curiosity almost guarantees someone will plug it in), or an email with an intriguing subject line that leads to a malware-infected website when clicked. Baiting relies on our innate curiosity, the impulse to explore things that seem out of the ordinary, even if it goes against better judgment.

Real-World Example: A Social Engineering Attack on a Small Business

Let's paint a picture—a seemingly ordinary day at a small accounting firm. The team is busy, juggling deadlines and tax forms, with just enough stress in the air to keep everyone on edge. It's precisely the kind of environment where social engineering thrives: fast-paced, overstretched, and vulnerable to the unexpected.

Enter the hacker, stage left, in the form of a professional-sounding phone call. The caller introduces themselves as a representative from the firm's trusted software provider. Their tone is confident, calm, and just urgent enough to grab attention. There's a critical update required, they explain, and the process simply can't be completed without the employee's login credentials. It's all very routine, very plausible—nothing to raise alarm bells.

The employee, eager not to cause delays or risk holding up the firm's workflow, obligingly provides the requested information. After all, who wants to be the one to slow everything down? But instead of securing the system, as promised, the caller (spoiler alert: a hacker) now has full access to the firm's digital vault.

Within hours, chaos quietly unfolds. Sensitive client data is compromised, financial records are siphoned off, and the firm finds itself staring down not only significant financial loss but also a reputational crisis. Clients are notified, trust is shattered, and all this stems from a single moment of misplaced confidence—a wolf in sheep's clothing pretending to be a trusted vendor.

This isn't a horror story; it's a cautionary tale. Social engineering attacks like this don't rely on brute force or technical wizardry. They rely on exploiting human nature—the instinct to trust, to cooperate, and to keep things running smoothly. And while the scenario may feel alarmingly simple, the fallout is anything but. It's a stark reminder that vigilance, even in the most mundane of moments, is key to keeping your business safe.

How to Defend Against Social Engineering

Social engineering might sound scary, but there are effective ways to protect your business from falling victim to these psychological tricks. Here are some key defenses:

1. **Verify Identities**: If someone contacts you out of the blue asking for sensitive information, always take a moment to verify who they are. A quick call back to the official contact number or a cross-check with a supervisor can prevent a lot of headaches.

2. **Question Unusual Requests**: Encourage your team to trust their instincts. If a request feels odd, out of character, or unusually urgent, it's worth a second look. Cybercriminals often rely on people acting before thinking, so a little skepticism can go a long way.

3. **Educate Employees on Red Flags**: Regular training sessions on social engineering tactics can make a world of difference. When employees know what to look out for—whether it's a suspicious email, a panicked caller, or a "lost" USB drive—they're less likely to be caught off guard.

4. **Create a Reporting Culture**: Encourage employees to report any unusual interactions, even if they're not sure it's a threat. A strong reporting culture helps everyone stay vig-

ilant, and it sends a message that cybersecurity is a shared responsibility.

Social engineering works because it's designed to feel natural, not threatening. But with a bit of awareness and a few preventive measures, you can close the door on these psychological manipulations. Remember, it's not about being suspicious of everyone, but about creating a healthy level of skepticism that keeps your business safe. When employees know what to watch for and feel empowered to question, they become your first line of defense against social engineering attacks.

Phishing: The Bait Hackers Use

What is Phishing?

Phishing is the digital equivalent of the oldest scam in the book, and for good reason—it works astonishingly well. At its heart, phishing is a type of social engineering that involves cybercriminals casting out their "bait" in the form of fake emails, text messages, or other seemingly innocuous communications. Instead of a fishing line dipped in a lake, they're dropping their hooks into your inbox, hoping someone will take the bait. And when someone does, it's not a trout they're reeling in, but a trove of sensitive information—passwords, credit card numbers, social security details, you name it.

Phishing works because it feels real. These messages are crafted to look like they're from someone you trust: your bank, a vendor you work with, or even your boss. The tone? Urgent. The language? Designed to make you act without thinking. "Update your account information immediately to avoid suspension!" or "Payment over-due—click here now!" It's the kind of high-stakes tone that gets your heart racing and your mouse clicking before your brain can intervene.

And that's the beauty—and the horror—of phishing. It preys on our instincts to react quickly in the face of urgency. Click that link, download that attachment, follow that seemingly legitimate instruc-tion, and suddenly you've handed over the keys to your digital king-dom. It's a masterful con, wrapped in everyday communication, and it can catch even the savviest of us off guard.

So, the next time an email makes your pulse quicken with its de-mands for immediate action, take a moment. Ask yourself: *Does this feel right?* Because, as with all good fishing trips, it's not the bait you need to watch—it's the hook.

Types of Phishing Attacks

Phishing comes in many forms, each with its own spin on the classic scam. Let's break down a few common types:

- **Email Phishing**: This is the most traditional form of phish-ing, where hackers send out fake emails pretending to be from trusted organizations like banks, delivery services, or big-name companies. These emails often contain a link or an attachment and may ask for sensitive information like lo-gin credentials, payment details, or personal data. The goal? Trick you into clicking and handing over valuable informa-tion without a second thought.

- **Spear Phishing**: If regular phishing is like casting a wide net, *spear phishing* is a targeted strike. Instead of sending out generic emails to thousands of people, hackers use spear phishing to focus on a specific individual or organization, often tailoring the message with personal details to make it more convincing. Imagine receiving an email that looks like it's from your boss, addressing you by name and referencing a recent project. The personalized touch makes it harder to spot as a scam, making you more likely to fall for it.

- **Smishing and Vishing**: Not all phishing happens over email. *Smishing* (SMS phishing) and *vishing* (voice phishing) use text messages and phone calls to deliver the bait. You might get a text message saying, "Your bank account has been compromised. Click this link to verify your identity." Or, you could receive a phone call from someone pretending to be tech support, claiming there's an issue with your computer and asking for remote access. These tactics are just as dangerous as email phishing and can be even more convincing because they're unexpected.

Common Phishing Tactics and Red Flags

The good news? Phishing attacks often come with telltale signs. Once you know what to look for, spotting these scams becomes a lot easier. Here are a few red flags that can help you steer clear of phishing attempts:

- **Misspelled Email Addresses**: Take a closer look at the sender's email address. A lot of phishing emails come from addresses that look almost right but aren't quite there—like cust0mer@amaz0n-support.com instead of customer@amazon.com. The differences are subtle, but they're often enough to reveal a fake.

- **Urgent or Threatening Language**: Phishing emails thrive on urgency. If you receive a message that says, "Immediate action required!" or "Your account will be deactivated!" it's worth a second look. Hackers use urgency to bypass your usual caution, hoping you'll act before thinking.

- **Requests for Sensitive Information**: A legitimate company will almost never ask for sensitive information (like passwords, social security numbers, or payment details) via email, text, or phone. If you're being asked to share this kind of information, it's likely a scam.

- **Suspicious Links**: Hover over any link in the email before clicking. Phishing links often lead to websites that look official but have slightly off URLs. If a link doesn't look like it's taking you to the company's actual website, it's best to avoid it altogether.

Here are a few classic examples of phishing messages you might encounter:

- A fake email from your bank saying, "Your account has been compromised. Click here to reset your password."

- A message claiming, "Your payment is overdue. Please confirm your payment details immediately."

- An email from "IT Support" asking you to verify your login credentials to keep your account secure.

Case Study: A Phishing Attack on a Small Business

Picture this: Jane, the hardworking owner of a small marketing agency, is juggling deadlines and client calls when an email pops into her inbox. It appears to be from her bank—slick, professional, with the bank's logo and just the right touch of official-sounding urgency. The email reads: "Dear Jane, we've detected unusual activity on your account. Please click the link below to verify your information and prevent your account from being suspended."

Now, Jane isn't someone who's easily rattled, but the thought of her account being suspended? That's enough to grab her attention. Feeling the pressure, she clicks the link, which whisks her away to what looks like her bank's website. It's all there—the familiar colors, the logo, the reassuring tone of professionalism. Without a second thought, she enters her username and password, believing she's safeguarding her account.

But here's the twist: the website isn't her bank's. It's a clever counterfeit set up by hackers. The moment Jane submits her credentials, they have exactly what they need. Minutes later, they're inside her real bank account, siphoning off her hard-earned funds faster than you can say "cybercrime." In the blink of an eye, Jane's business finances are compromised—all because of one innocent-looking email and a single click.

This isn't a far-fetched cautionary tale; it's an everyday reality. Phishing attacks like this one are disturbingly effective because they create a sense of urgency, a ticking clock that makes even cautious people throw their usual skepticism out the window. The lesson? If something feels urgent and too good—or too bad—to be true, pause, breathe, and double-check. Because, as Jane learned the hard way, not every link leads where it promises to go.

How to Protect Against Phishing

The best way to defend against phishing is awareness. Here are some tips to help you and your team recognize and avoid these sneaky scams:

1. **Double-Check the Sender's Address**: Always examine the sender's email address carefully, especially if the message seems out of the ordinary. Even a slight misspelling can indicate a phishing attempt.

2. **Don't Click on Unsolicited Links**: If you receive an email or text asking you to click a link, pause. Hover over the link to see the actual URL, and if anything seems off, it's safer not to click. Visit the official website directly instead.

3. **Verify Requests by Contacting the Source Directly**: If you get a message from someone claiming to be from your bank, vendor, or colleague asking for sensitive information, take a moment to confirm it's legitimate. Contact the person or organization directly through a verified contact method (not the contact details in the suspicious message).

4. **Enable Multi-Factor Authentication (MFA)**: MFA adds an extra layer of security, requiring a second form of verification (like a text code) in addition to your password. Even if hackers get your password, they can't access your account without that second step.

5. **Educate Your Team**: Phishing awareness training can go a long way in preventing attacks. Ensure that everyone in your business knows what phishing looks like and understands the importance of reporting suspicious messages.

Phishing is one of the most common and effective tactics in a cybercriminal's arsenal, but by staying vigilant and educating your team, you can minimize the risk. Remember, hackers rely on people's reflexes and instincts, hoping for a momentary lapse in judgment. By taking a moment to double-check and question anything that seems out of place, you're already one step ahead in protecting your business from the bait hackers are casting.

Malware: Viruses, Worms, and Trojans, Oh My!

Step right up and behold the wily villains of the digital world, where cyber miscreants skulk in the shadows, poised to pounce on unsuspecting systems. It's a bit like Dorothy's journey in *The Wizard of Oz*—only instead of lions, tigers, and bears, we've got viruses, worms, and Trojans. And believe me, "Oh my!" is a perfectly reasonable reaction when you realize what these troublemakers are capable of.

This malware menagerie comes equipped with its own nasty bag of tricks. Viruses, for instance, behave just like their biological counterparts, spreading rapidly and causing havoc as they go. Worms are their ambitious cousins, slithering through networks with alarming speed, while Trojans—well, they're the sneaky ones. They show up looking harmless, like a gift-wrapped surprise, only to reveal themselves as something altogether more sinister once they've wormed their way i nside.

These are not your garden-variety glitches; they're the digital equivalent of vandals, saboteurs, and thieves, capable of turning your serene digital landscape into a chaotic nightmare. But don't worry—we're here to pull back the curtain on this rogues' gallery so you'll know exactly what you're up against. By the end of this journey, you'll be better equipped to spot the threats, thwart their antics, and keep your systems safer than Dorothy's ruby slippers. Let's dive into the malware zoo, shall we?

What is Malware?

Let's begin with the basics. Malware—short for "malicious software"—is essentially the digital equivalent of an uninvited houseguest who not only overstays their welcome but also trashes the place, rifles through your drawers, and walks off with the silverware. Designed with a singular purpose—to harm, disrupt, or infiltrate without your permission—malware is the mischief-maker of the cyber world, and it's alarmingly good at its job.

Picture this: you're peacefully working away, blissfully unaware that somewhere in the hidden recesses of your computer, malware is quietly doing its thing. Maybe it's slowing down your system, mucking about with your files, or, worse still, making off with your most sensitive data. For cybercriminals, malware isn't just a tool—it's a Swiss Army knife, offering a variety of ways to cause trouble.

And trouble, it turns out, comes in many forms. Malware has a whole repertoire of tricks, from viruses that spread chaos to worms that wriggle their way through networks, and Trojans that sneak in under the guise of something helpful. Each type has its own brand of sneakiness, but they all share a common goal: to get in, cause havoc, and leave you wondering what on earth just happened.

Malware may be the bane of the digital age, but understanding it is the first step in defending against it. So buckle up, because we're about to venture into the murky world of malicious software, where nothing is quite as harmless as it seems.

Different Types of Malware in Plain English

Now, let's break down the different types of malware you might encounter in the wild. Each type has its own "personality" and modus operandi, so understanding them is your first step toward keeping them at bay.

- **Viruses**: Just like a flu virus, a computer virus spreads by attaching itself to a host—in this case, a legitimate file. When that file is opened or shared, the virus spreads to other files or devices. It can slow down your system, corrupt files, or even cause programs to crash. Imagine opening a file, only to have it "infect" the rest of your system without your knowledge. Not exactly a friendly surprise.

- **Worms**: Worms are like viruses on a mission. They don't need a host file or any human interaction to spread; they replicate themselves and jump from computer to computer on their own. Worms are particularly dangerous because they can spread rapidly across a network, using up resources and potentially causing massive disruptions. Think of them as digital pests, burrowing through your systems and leaving a trail of chaos.

- **Trojans**: This type of malware takes a page from ancient Greek mythology. Much like the infamous Trojan Horse, a Trojan disguises itself as something helpful or legitimate—like a game or a free software download. Once it's inside, though, it reveals its true intentions, often opening the door for other types of malware or even allowing hackers remote access to your device. The lesson? Not everything that looks helpful on the surface can be trusted.

- **Spyware**: As the name suggests, spyware is the cyber equivalent of a stalker. It secretly monitors your activities, often tracking everything from the websites you visit to your keystrokes, in order to steal sensitive information like login credentials or credit card numbers. Spyware doesn't announce its presence; it lurks quietly in the background, observing and reporting your every move.

- **Adware**: Adware's primary purpose is to bombard you with advertisements. While it's not always harmful, it can be incredibly annoying, popping up unwanted ads on your screen and potentially slowing down your device. The risk with adware is that those ads may lead you to other malicious websites, turning a nuisance into something far more dangerous.

How Malware Infects Devices

You might be wondering, "How does this stuff even get onto my devices?" Malware has a way of sneaking in through everyday activities, often in ways that seem innocent enough. Here are some common infection methods:

- **Infected Attachments**: That email attachment that looks like an invoice? It could be carrying a virus. Hackers often disguise malware as legitimate attachments to trick people into downloading them.

- **Malicious Websites**: Visiting certain websites (especially ones offering free downloads or streaming) can expose you to malware. Sometimes, just loading a site is enough for malware to start downloading.

- **Infected USB Drives**: Ever plug in a USB drive without knowing where it came from? An infected drive can transfer malware to your system as soon as it's connected.

In other words, malware often relies on us making a small mistake—like clicking on the wrong link or downloading a file we thought was safe.

A Malware Infection in a Small Business

Imagine this: a small, overworked accounting firm stumbles across a free software program that promises to revolutionize their workflow. It's shiny, professional-looking, and has glowing reviews plastered across a slick website. "Why not?" they think, eager to shave a few minutes off their already hectic day.

But here's where our story takes a turn for the worse. Unbeknownst to the unsuspecting accountants, this seemingly helpful software comes with an unwanted passenger—a Trojan, stealthily bundled in like the world's least charming surprise gift. At first, everything seems fine. The software works as advertised, humming along in the background.

Then, the cracks begin to show. Client data, the lifeblood of their business, starts leaking into the digital ether, landing in the hands of who-knows-where. Mysterious account logins pop up like unwelcome guests, and before they know it, the firm is grappling with a full-blown crisis. Systems need scrubbing, passwords need resetting, and panicked calls to clients become the order of the day. But the real kicker? The financial and reputational damage proves even harder to repair than the technical mess.

The worst part? This scenario isn't some far-fetched, doom-and-gloom tale—it's a remarkably common story. Small businesses, often operating on tight budgets and even tighter schedules, are prime targets for traps like these. That shiny, free program? It's bait, plain and simple, and the hook is sharper than you'd think.

The lesson here is clear: in the world of software, free isn't always a bargain. A little caution can save a lot of heartache, and sometimes, the best investment you can make is in security—both for your systems and your peace of mind.

Steps to Prevent Malware Infection

Now that you know what malware is and how it sneaks in, let's talk about some simple ways to keep it out. Here's how you can bolster your defenses against these digital intruders:

1. **Download Software Only from Trusted Sources**: If a program isn't from a verified, reputable source, think twice before downloading it. This is especially important for free software, which can sometimes come bundled with hidden malware.

2. **Use Up-to-Date Antivirus Software**: Antivirus programs act as your first line of defense, scanning for known malware and blocking potential threats. Just make sure it's updated regularly so it can keep up with new types of malware.

3. **Educate Employees About Suspicious Links and Downloads**: The best defense is a well-informed team. Make sure everyone knows not to click on links or download attachments from unknown or untrusted sources. A quick reminder can save a lot of trouble.

Remember, in the digital world, caution is key. Malware relies on trickery and misdirection, but by staying vigilant and following these simple steps, you can keep your business safe from unwelcome guests. Just like locking your doors at night, good cybersecurity habits can make all the difference in protecting what you've worked so hard to build.

Ransomware: When Your Data Gets Held for Ransom

If the world of cyber threats were a genre of film, ransomware would be the quintessential kidnapping thriller—but with a distinctly modern twist. Forget shadowy figures in trench coats; here, the hostages are your files, and the kidnappers are hackers who've locked them away in a digital vault. Their demands? Cold, hard cryptocurrency, delivered quickly, or your precious data gets it.

Ransomware is a particularly devious type of malicious software. It sneaks into your system, encrypts your files, and then leaves you staring at a grim ultimatum: pay up, or kiss your information goodbye. It's like waking up to find your office safe bolted shut, with a ransom note taped to the front, except the "safe" in this case is your computer/network server/cloud servers/even off-site backups, and the ransom note appears as a menacing message on your screen.

The whole affair is as chilling as it is clever. Hackers know that the thought of losing important data—financial records, client files, irreplaceable intellectual property—will send most people into a panic. That panic, in turn, leads some victims to pay the ransom in a desperate bid to restore their systems. Of course, the ransom is almost always requested in cryptocurrency, ensuring the hackers remain comfortably anonymous while you scramble to pick up the pieces.

Ransomware is more than just a nuisance; it's a reminder of just how valuable your data is—to you and, unfortunately, to those who would exploit it. It's a digital drama you'd rather not star in, but understanding the plot is the first step in making sure you're not cast in the role of the victim.

How Ransomware Works

The ransomware drama typically begins innocently enough—or at least it seems that way. It might sneak into your system through a phishing email that looks like it's from a trusted source, or it might sidle in through a vulnerability in software that hasn't been updated since, well, who knows when. Either way, the malicious software sets up shop quietly at first, biding its time like a cat ready to pounce.

Then, suddenly, all hell breaks loose. The ransomware spreads through your network, encrypting files with the efficiency of a librarian who's decided every book belongs behind a locked glass case. Files you once accessed with ease are now as inaccessible as a medieval treasure chest without the key. And that's when the pièce de résistance appears: a jarring message splashed across your screen, cheerfully informing you, *"Your files are encrypted. Pay $10,000 (frequently even more) in Bitcoin within 72 hours, or they'll be deleted forever."* Sometimes, they'll take it to the next level and threaten to make all of your, (and that of your clients) information publicly available.

It's a digital hostage note, complete with a ticking clock. The countdown isn't just for drama; it's a calculated scare tactic, designed to send your stress levels skyrocketing. The pressure mounts, and businesses—terrified of losing critical data—often scramble to pay the ransom, desperately hoping the hackers will play nice and restore access.

Of course, there's no guarantee they will. After all, you're dealing with cybercriminals, not a customer service department. But that's precisely what makes ransomware so fiendishly effective: it weaponizes fear, urgency, and the importance of your data, leaving you in a high-stakes guessing game with your business on the line.

The Growing Threat of Ransomware Attacks on Small Businesses

It's tempting to think of ransomware as the sort of thing that happens to massive corporations with sprawling networks and deep pockets. After all, why would cybercriminals bother with the local florist or a family-run accounting firm when they could go after a Fortune 500 company? But here's the sobering truth: small businesses are increasingly the targets of ransomware attacks. Why? Because, to cybercriminals, they're the digital equivalent of an unlocked car with the keys still in the ignition—quick, easy, and highly rewarding.

For small businesses, the stakes couldn't be higher. A ransomware attack can grind operations to a halt, leaving employees twiddling their thumbs while encrypted files mock you from behind their digital padlocks. Productivity screeches to a standstill, revenue vanishes, and expenses for recovery skyrocket. And the kicker? Without proper backups or defenses in place, the business's very survival can hang in the balance.

It's not just about losing a few days of work or paying a hefty ransom; it's the ripple effect. Clients lose trust, reputations take a hit, and in some cases, the business itself might never fully recover. For cybercriminals, small businesses are prime targets because they know many simply don't have the resources to fight back. It's a ruthless game, and for small businesses, the stakes couldn't feel more personal.

A Ransomware Attack on a Small Business

Picture this: a local accounting firm in the thick of tax season. The phones are ringing, the coffee is flowing, and the office is buzzing with the collective hum of calculators and keyboards. It's business as usual—until it isn't.

One day, an employee clicks on a link in an email that looks harmless enough. It appears to be from a trusted client, with the subject line, *"Urgent: Tax Documents Attached."* But instead of leading to a helpful PDF, the link unleashes a digital Pandora's box. Within minutes, files across the company's server begin to lock up tighter than a Victorian corset.

The damage is swift and merciless. Financial records? Encrypted. Client information? Encrypted. The tax preparation software that's the lifeblood of their operation? You guessed it—encrypted. Then comes the pièce de résistance: a ransom message glowing ominously on the screen. The demand? $20,000 in cryptocurrency. The promise? A decryption key to restore access—if the hackers can be trusted, of course.

With no recent backups to fall back on (even if you have recent backups, even the backups may have also been encrypted as well) and a growing chorus of anxious clients demanding updates, the firm faces an impossible decision. Should they scrape together the ransom money and pray the hackers keep their word, or write off the data as a loss and risk shutting their doors for good?

It's a nightmare scenario, but one that's all too real for countless small businesses. This isn't just a tale of woe; it's a cautionary reminder of how critical ransomware preparedness is. Because in the world of cybersecurity, being caught unprepared can mean the difference between business as usual and business no more.

How to Protect Against Ransomware

The good news? There are ways to protect yourself. Here are some practical steps to defend against ransomware attacks:

- **Regularly Back Up Data and Store Backups Offline**: This step is like having a safety net. Even if ransomware infects your systems, having recent backups stored offline ensures you won't lose everything.

- **Keep Software Up-to-Date**: Cybercriminals love exploiting outdated software with known vulnerabilities. By keeping your systems patched and updated, you reduce entry points for ransomware.

- **Implement Strong Email Filtering**: Many ransomware attacks start with phishing emails. Having strong email filters can catch suspicious messages before they even reach your inbox, reducing the chances of accidental clicks.

What to Do if You're Hit by Ransomware

If the worst happens and you find yourself in the grip of a ransomware attack, take a deep breath. Here's a quick list of actions to take:

1. **Don't Panic**: Staying calm is essential for handling the situation effectively.

2. **Isolate Affected Systems**: Disconnect infected devices

from the network to prevent the ransomware from spreading further.

3. **Report the Incident**: Notify relevant authorities and possibly your cyber insurance provider if you have one.

4. **Seek Professional Help**: Contact cybersecurity professionals who specialize in ransomware recovery. They may be able to help you restore files without paying the ransom.

Ransomware is a formidable adversary, but with awareness and the right preventive measures, you can reduce the risk of an attack and protect your business from becoming just another statistic.

Wrap-Up: Recognizing the Hacker's Playbook

As we bring this chapter to a close, let's step back for a moment to admire the hacker's playbook—not out of respect, of course, but to better understand the common threads that make their schemes so maddeningly effective. From social engineering to phishing and ransomware, there's a certain artistry (albeit of the criminal variety) in how these strategies work.

At their core, hackers are master manipulators of human behavior. They don't bash through digital walls like cyber-thugs; they knock politely, dressed as someone you trust, and wait for you to let them in. Whether it's a cleverly disguised email, a fake login page, or a ransom note that feels like it came straight out of a heist movie, their methods are designed to make you act first and think later.

Another theme? They're opportunists, plain and simple. Hackers thrive on businesses that aren't prepared. Outdated software, weak passwords, and a lack of backups are like neon signs that say, *"Wel-*

come! Please rob us." They count on the chaos, panic, and vulnerability that unpreparedness brings, using it to exploit systems and people alike.

But here's the good news: by understanding these overarching tactics, you're already a step ahead. Recognizing the patterns behind these digital intrusions helps you see them for what they are—manipulations designed to trick you into opening the door. And once you start spotting the signs, you'll find that keeping those doors firmly shut becomes much, much easier.

The Importance of Awareness and Vigilance

Here's the crux of it: awareness is your first and most formidable line of defense. Knowledge, as the old saying goes, isn't just power—it's protection. When you're clued in on the tactics hackers love to deploy, you start seeing the digital world with a sharper eye, equipped with a kind of skepticism that would make even the most cynical detective proud.

Now, let's be clear—this isn't about spiraling into paranoia every time an email lands in your inbox. It's about cultivating a sense of vigilance. A little wariness, a moment to ask, *"Does this look right?"* or *"Should I really click that?"* can make all the difference in keeping your business safe. Hackers count on people moving too quickly, acting without questioning, and leaving the metaphorical barn door wide open. But vigilance? Vigilance slams that door shut.

The truth is, hackers thrive in complacency. They're like weeds in an untended garden, flourishing wherever there's a lack of care or attention. Awareness disrupts that cozy little ecosystem, stripping them of their greatest advantage. By staying alert, asking questions, and thinking twice before you click, you're not just protecting your data—you're sending a clear message to those digital miscreants: *Not today, thank you very much.*

Building a Culture of Cybersecurity Awareness

Now that you're armed with this understanding, let's pause for a moment and consider the possibilities. Imagine a workplace where cybersecurity isn't seen as some mystical realm reserved for the IT department but instead as a shared responsibility, woven into the very fabric of your business. Picture a team where everyone—not just the tech-savvy ones—has the knowledge and confidence to recognize threats and act decisively.

This is where you step in. Not just as the leader of your business, but as the spark for a cultural shift. By educating your employees, encouraging candid conversations about cybersecurity (even the embarrassing *"Oops, I clicked on that link"* ones), and fostering an environment where vigilance is second nature, you're doing more than raising awareness—you're building a collective shield. Every team member who's aware and alert adds another layer to your defense, turning your business into a far less inviting target for cyber mischief-makers.

With each chapter, we're peeling back the layers of cybersecurity, revealing the practical steps you can take to strengthen your defenses. In the next chapter, we'll roll up our sleeves and get into the nitty-gritty of specific security measures you can implement. It's all part of the journey toward a safer, more resilient business.

So, take a deep breath and pat yourself on the back. You're not just learning—you're actively building a fortress, one sturdy, sensible brick at a time. And trust me, when it comes to protecting your business, that's a foundation worth celebrating.

Conclusion: Ready for the Next Chapter – Building Defenses

Now that we've spent some time rummaging through the hacker's playbook, you're no longer the wide-eyed bystander in the cybersecurity drama. You're the informed, slightly skeptical protagonist who knows a trick or two. But here's the thing: recognizing the traps is only half the battle. The real triumph comes in building defenses so solid that hackers take one look and decide to slink off elsewhere.

In the next chapter, we'll roll up our sleeves and dive into the nitty-gritty of actively defending against the threats we've explored. From phishing scams to ransomware—and all those delightfully underhanded schemes in between—you'll learn how to set up essential cybersecurity measures to keep your business safe.

And no, this doesn't require a PhD in computer science or a sudden urge to learn binary. What it does take is a willingness to get stuck in, a dash of commitment, and a commitment to protecting the business you've worked so hard to build. Think of it as fortifying your digital castle—strong walls, a sturdy gate, and maybe a few metaphorical alligators in the moat for good measure. Let's get to work!

Encouragement to Stay Informed

As you journey through the ever-evolving world of cybersecurity, here's a comforting truth: this isn't a sprint to some elusive finish line; it's more like tending a garden. It's a continuous process—watering here, weeding there, and staying just attentive enough to keep things flourishing. A quick glance at an article, a podcast episode during your morning coffee, or even a passing update from your IT team can go a long way in keeping you ahead of those pesky digital pests.

You're already doing something remarkable by taking these first steps toward securing your business. With every bit of knowledge you gain, you're not just patching holes in your defenses—you're building a fortress, brick by digital brick. You've got this! With the right tools, a touch of vigilance, and a growing awareness of the digital landscape, you're more than capable of steering your business safely through the wilds of cyberspace.

Now, it's time to roll up those sleeves and dive into the practical side of things. Let's get to work building defenses that would make even the savviest hacker think twice about knocking on your virtual door. Onward!

CYBER HYGIENE

KEEP IT CLEAN, KEEP IT SAFE

Introduction to Cyber Hygiene

Welcome to the marvelous world of cyber hygiene—a place where a handful of simple habits can mean the difference between your business humming along happily and the digital equivalent of catching a nasty case of the flu. Think of cyber hygiene as you would brushing your teeth or washing your hands. These seemingly mundane routines don't just keep you feeling fresh—they fend off cavities, germs, and those dreadful visits to the dentist.

The same principle applies to your business's digital health. A few well-practiced habits can keep your systems clean, functional, and blissfully free of cyber "germs" (or, in our case, hackers and malware). These practices aren't about grand gestures or flashy technology—they're about consistency. Much like remembering to floss, they're things you do regularly to avoid much bigger headaches later on.

So grab your metaphorical toothpaste and toothbrush, and let's dive into the world of cyber hygiene. It's not glamorous, but trust me, it's essential. And just like good personal hygiene, it'll leave you feeling a whole lot better knowing you've taken care of business where it really counts.

The Benefits of Practicing Good Cyber Hygiene

Why, you might ask, should you give two hoots about cyber hygiene? Well, let me put it this way: it's like eating your vegetables. It may not seem thrilling at first, but it's one of the smartest things you can do to ward off trouble down the road. Think of cyber hygiene as your business's proactive shield—your digital spinach, if you will—reducing the likelihood of expensive, time-consuming, and trust-eroding disasters like data breaches, ransomware attacks, and accidental data loss.

When your digital house is tidy and in good order, cybercriminals find it much harder to get a foothold. A well-maintained system doesn't just fend off potential threats; it also cuts down on downtime and spares you the steep costs of reactive fixes. Let's be honest—patching up the aftermath of a cyber attack is a bit like trying to fix a leaky roof during a thunderstorm. It's messy, it's expensive, and you'll wish you'd taken care of it sooner.

Investing a little time in good cyber hygiene now is like laying out a preventative safety net. It saves you a world of headaches, a pile of cash, and perhaps most importantly, preserves that all-important trust your customers have in you. Because, in the grand scheme of things, a tidy digital space doesn't just make life easier—it keeps your business humming along beautifully.

How Small Steps Add Up to Big Protection

Here's the reassuring truth about cyber hygiene: you don't need a degree in computer science or a vault full of cash to get it right. Much like washing your hands or brushing your teeth, it's the small, regular actions that make all the difference. A strong password here, a timely software update there—it all adds up. And for those of us who might feel a bit overwhelmed by the digital realm, this is particularly good news. You don't have to rebuild your entire operation overnight; just a few sensible habits can provide surprisingly sturdy protection.

Think of it this way: each small step you take—whether it's running updates, backing up files, or teaching your team how to spot phishing emails—is like adding another brick to your cybersecurity wall. It's not flashy, and it's certainly not rocket science, but it's effective. And best of all? These practices aren't just for the tech-savvy elite. They're designed to be straightforward, practical, and manageable for any business.

In this chapter, we'll break down the essentials of cyber hygiene into bite-sized, actionable tips. Consider it your digital cleaning checklist—a way to keep your virtual workspace as spotless and secure as your physical one. By the end, you'll have a tidy toolkit of habits that will make cybersecurity feel less like a chore and more like second nature. So grab your metaphorical sponge, and let's start scrubbing down those digital surfaces!

Password Protection 101

Ah, passwords—the unsung heroes of the digital age. Those seemingly innocuous strings of letters, numbers, and symbols are the last line of defense between you and the cybercriminals skulking in the virtual shadows. But let's be honest—if your go-to password is "Password123" or (brace yourself) "123456," we need to have a chat. Weak, predictable passwords are the digital equivalent of leaving your front door not just unlocked but propped wide open with a sign saying, *"Welcome! Please help yourself to everything inside!"*

Fear not, though, because we're here to dig into why good password habits are essential, how to make your passwords hacker-proof, and how to manage it all without driving yourself mad in the process.

Why "Password123" is Asking for Trouble

Let's face it—using something simple is tempting. Who has the time, or the memory, to juggle a dozen different complex passwords? But here's the rub: hackers are banking on you taking the easy route. Common passwords like "Password123" or the ever-lovable "qwerty" are so popular they're basically hacker catnip. These gems are the first things cybercriminals try because they know countless people rely on these easy-to-remember combinations.

To put it bluntly, if your password can be cracked faster than it takes to make a cup of coffee—or worse, if it could be guessed by a moderately curious toddler—it's time for a serious upgrade. Yes, it's inconvenient to think up and remember stronger passwords, but trust me, it's a small price to pay for keeping your digital fortress intact.

Stick around, because we're about to turn those digital skeleton keys into locks worthy of Fort Knox. You'll learn how to make passwords that hackers hate, keep them all organized, and still get on with your day without a single post-it note stuck to your monitor. Let's dive in!

Tips for Creating Strong Passwords

Crafting a strong password isn't exactly rocket science, but it does require a pinch of thought and a dash of creativity. It's less about building a digital masterpiece and more about ensuring that any hackers attempting to break in feel like they're trying to open Fort Knox with a plastic spork. Here's how to make passwords that are as infuriatingly uncrackable as possible:

Length and Complexity: Think big. Aim for at least 12 characters and toss in a delightful mix of uppercase and lowercase letters, numbers, and symbols. The longer and messier, the better. This isn't the time for minimalism—go ahead and make it a fortress.

Avoiding Personal Information: Sure, your childhood pet's name or your favorite football team might feel personal, but guess what? Hackers know how to scrape your social media for those little nuggets of information. If it's been posted, tweeted, or grumbled about online, consider it off-limits for password duty.

Using Unique Passwords for Each Account: One password for everything may seem efficient, but it's the digital equivalent of using the same key for your house, car, and bank vault. If that key gets copied, well, you see the problem. Each account deserves its own password to prevent a chain reaction of disasters.

Why Password Managers are Your Best Friend

If the idea of juggling dozens of unique, complex passwords has you sweating, let me introduce your new best friend: the password manager. These handy tools do all the heavy lifting for you. They'll generate strong, unique passwords for every account and store them securely, leaving you with just one master password to remember.

It's like having a butler for your digital life—only instead of carrying trays, they're carrying the keys to your online kingdom. Options like LastPass, Dashlane, and Bitwarden offer user-friendly features that make managing passwords a breeze. With a password manager in your corner, you can keep your digital fortress secure without resorting to scribbling passwords on sticky notes. Speaking of which...

Common Password Mistakes to Avoid

Even with the best of intentions, it's all too easy to stumble into a few password pitfalls. Here's a quick rundown of what *not* to do:

- **Using Simple Passwords**: If your password could be cracked by a mildly curious hamster, it's time to rethink your strategy. Short and simple might be easy to remember, but it's also easy to hack.

- **Writing Passwords on Sticky Notes**: Let's face it, we've all seen them—little yellow reminders stuck to monitors with "secure" passwords scribbled on them. It's convenient, sure, but also a glaring security risk. If you must write them down, keep the list somewhere secure, like a locked drawer. Not writing them down at all is an even better option. Better yet, use a password manager.

- **Sharing Passwords Over Email or Chat**: Sending a password to a coworker via email or messaging might seem harmless, but those channels can be intercepted. If you need to share a password, use a secure method or a password manager's sharing feature.

By following these tips, you'll create a solid foundation for password security that'll leave hackers scratching their heads—and hopefully, moving on to easier targets. Remember, your password is your first line of defense, so make it count. Think of it as the digital equivalent of a sturdy, unbreakable lock. Or better yet, the moat and drawbridge to your online castle.

Multi-Factor Authentication: Your Digital Bouncer

You've probably come across the term Multi-Factor Authentication (MFA) before. If not, allow me to introduce you to what is essentially the bouncer of the cybersecurity world. Think of MFA as an added security guard stationed outside the door to your digital assets, clipboard in hand, ensuring only the right people get in. Because, let's face it, passwords—no matter how cleverly crafted—aren't foolproof. Cybercriminals are devious creatures, always sniffing around for ways to crack even the strongest defenses. That's where MFA steps in, adding an extra hurdle that makes it exponentially harder for intruders to worm their way in.

Let's unpack why MFA is such a game-changer and how you can implement it without tearing your hair out.

What is Multi-Factor Authentication (MFA)?

In plain terms, Multi-Factor Authentication, or MFA, is the cybersecurity equivalent of asking for not just one form of ID but two. It's an extra layer of security designed to ensure that the person trying to access your account is actually you. Instead of relying solely on a password—which, let's be honest, might have been guessed by your nosy neighbor—MFA adds a second (and sometimes third) step to the login process.

That extra step could be a code sent to your phone, a fingerprint scan, or a one-time passcode generated by an app. The magic lies in combining different types of verification:

1. **Something you know** (your password).

2. **Something you have** (a phone or device).

3. **Something you are** (a fingerprint or facial recognition).

Even if a hacker gets their grubby digital hands on your password, they'll still hit a brick wall without that second piece of evidence. It's like putting a deadbolt on top of a lock—it won't stop someone determined enough to break down the door, but it'll make their job significantly harder. And in the world of cybersecurity, every extra hurdle counts.

So, whether you're protecting your email, banking information, or a treasured collection of cat memes, MFA is your trusty sidekick in keeping cyber villains at bay. It's simple to set up, powerful in its effectiveness, and a downright no-brainer for anyone serious about safeguarding their digital life. Let's get started and make your accounts as impenetrable as a fortress.

How MFA Works: An Analogy

Imagine trying to access a bank vault—not just any vault, but one of those movie-style, over-the-top fortresses of security. Knowing the door code isn't enough. You also need a key card, and maybe even a fingerprint scan, to get inside. That's the essence of MFA, or Multi-Factor Authentication. It's like having not one but two locks on the same door. The first lock is your password—important, yes, but hardly invincible. The second lock? That's MFA, the extra layer that ensures even if someone manages to "pick" the first, they're left standing helplessly in front of the second.

This simple but brilliant concept transforms a flimsy lock into a fortress, keeping your digital valuables much safer from prying eyes.

Different Types of MFA

Not all MFA methods are created equal, but they all aim to make life harder for hackers. Here's a whirlwind tour of the most common types, along with their pros and cons:

- **SMS Verification**: This classic method sends a one-time code to your phone via text every time you log in. You punch in the code alongside your password to prove it's really you. It's simple, it's widely available, and it's a good starting point. But beware—SMS verification isn't the most secure option. Phone numbers can be hijacked (it's called SIM swapping, and it's as annoying as it sounds), so while better than no MFA at all, it's not exactly Fort Knox.

- **Authenticator Apps**: Meet the slightly nerdier cousin of SMS verification. Apps like Google Authenticator, Authy, and Microsoft Authenticator generate time-sensitive codes that you enter alongside your password. These codes refresh every 30 seconds and aren't tied to your phone number, making them a stronger option. Think of it as an extra-secure stopwatch working in your favor.

- **Biometrics**: If you've ever unlocked your phone with your fingerprint or gazed lovingly into its camera for facial recognition, congratulations—you've dabbled in biometrics. This method uses something unique to you—your fingerprint, face, or even your voice—to verify your identity. Biometrics are growing in popularity because they're both convenient and devilishly difficult for hackers to replicate. Unless someone steals your face (a bit extreme, but hey, we're covering all bases), you're pretty safe.

Each method has its charm, so pick the one that suits your needs. Just remember that while all MFA is better than no MFA, some methods are more robust than others.

Benefits of MFA for Small Businesses

For small businesses, enabling MFA is the cybersecurity equivalent of adding a steel gate behind your front door. Hackers might still try to jiggle the lock, but once they realize they'd also need to climb over a spiked fence, they'll likely move on to easier pickings.

Here's the magic of MFA: even if a hacker gets hold of your password, they're still stuck. Without that second factor—whether it's a code, a scan, or a swipe—they're effectively locked out. This drastically reduces the risk of unauthorized access and spares you the nightmare of a data breach, financial loss, or the dreaded *"We regret to inform you..."* email to your clients.

For small businesses without the budget for a full-blown IT security team, MFA is a simple, cost-effective way to level up your defenses. It's not flashy or complicated—it's just smart. And in the ever-shifting sands of cybersecurity, that's worth its weight in gold.

How to Enable MFA on Common Platforms

If the idea of setting up Multi-Factor Authentication (MFA) feels daunting, take heart—it's easier than assembling flat-pack furniture, and far less frustrating. Most major platforms practically roll out the red carpet for you, offering simple, step-by-step guides to get MFA up and running. Here's a quick overview of where to start:

- **Email Accounts**: Gmail, Outlook, and other email platforms have MFA options baked right into their security settings. Pop into your account, look for something like "Security" or "Account Settings," and voilà—there should be an option for two-step verification or MFA. Follow the prompts, and you'll be adding a second layer of protection to your inbox in no time.

- **Cloud Storage**: For services like Dropbox, Google Drive, and OneDrive, MFA is your digital padlock. Head to the security settings, enable MFA, and enjoy the added peace of mind knowing your files are tucked safely behind a second layer of defense.

- **Financial Accounts**: If there's one place you *definitely* want MFA, it's anywhere your money lives. Banks and payment platforms like PayPal have MFA options that are usually found under "Security" or "Privacy" settings. Log in, follow the prompts, and sleep easier knowing that your finances are better protected.

What If You're Not Sure How?

If the tech jargon starts to blur together or you find yourself squinting at your screen, fear not. Most platforms have a help section with step-by-step instructions, often accompanied by cheerful little diagrams or screenshots. And if you're still feeling unsure, don't hesitate to ask a tech-savvy friend or someone on your team for help. MFA setup is one of those tasks where asking for assistance is far better than risking a half-done job.

Why MFA Matters

By enabling MFA, you're not just ticking a box—you're taking a meaningful step to safeguard your business. It's like adding a sturdy deadbolt to your digital doors. Cybercriminals might knock, but with MFA in place, they're far less likely to get in.

So, go ahead, take a few minutes to set it up. You'll be adding a layer of security that makes all the difference, and you'll have the added bonus of feeling like a cybersecurity pro. Plus, there's nothing quite as satisfying as knowing you've just made life that much harder for the hackers of the world.

Regular Updates and Patch Management

Let's be honest—keeping your software updated probably doesn't rank high on your list of thrilling tasks. It's not going to win any awards for glamour, but it's one of the most important habits you can adopt for cybersecurity. Think of each update as swapping out an old, wobbly lock on your door for a shiny new one that's virtually pick-proof. Hackers love nothing more than outdated software; it's their playground of vulnerabilities. Staying on top of updates and patches is like putting up a bold, unmistakable "Keep Out" sign for digital intruders.

Why Updating Software Matters

Every piece of software—whether it's your computer's operating system, the apps on your phone, or even the widgets running your smart fridge—has potential vulnerabilities. Developers, like industrious locksmiths, are constantly hunting for these weaknesses, releasing updates to patch them before hackers can take advantage.

When you skip an update, you're effectively leaving the door ajar and hoping no one notices. Maybe it's because you forgot, or maybe you hit "Remind Me Later" for the hundredth time—but whatever the reason, ignoring updates is like leaving a rusty old lock on your front door. Hackers see it and think, *Ah, what a delightful invitation!* Regular updates are one of the simplest, most effective ways to keep your defenses strong and your digital doors securely bolted.

Common Types of Updates and What They Do

Not all updates are cut from the same cloth. Let's break down the usual suspects:

- **Security Patches**: These are the unsung heroes of the update world. They fix specific security flaws that hackers love to exploit. Think of them as urgent repairs to a breach in the wall—essential, immediate, and non-negotiable. When a security patch comes through, treat it with the same seriousness as fixing a flat tire on a busy highway.

- **Feature Updates**: These updates are the shiny new toys in the software world. They often introduce new functionalities or improve usability. While their primary purpose is enhancing your experience, they frequently bundle in security improvements as well. It's like upgrading to a better lock that also happens to be easier to use.

- **Bug Fixes**: These deal with the little gremlins in the system—the odd hiccups and glitches that make software behave like it's having a bad day. While they might not scream "urgent," bug fixes often close off small vulnerabilities that hackers could exploit. Think of them as preventive maintenance, ensuring that minor issues don't snowball into major headaches.

Why You Shouldn't Ignore Updates

Each type of update plays a crucial role in keeping your systems secure and functional. Ignoring them is the digital equivalent of refusing to maintain your locks or alarm system. Sure, it might not cause a problem today or even tomorrow, but it's only a matter of time before something goes awry.

So the next time you see that "Update Available" notification, don't groan and postpone it. Instead, take a moment to appreciate the humble update for what it is: a steadfast defender of your digital fortress. And remember, every little patch and upgrade you apply is another step toward a safer, smoother-running business. It might not be glamorous, but it's definitely worth it.

A Cautionary Tale About Ignoring Updates

If you've ever thought, *Eh, I'll get to that update later,* let me share a tale that might make you rethink procrastination. A few years ago, a major cyberattack made headlines when hackers exploited an unpatched vulnerability in a widely used software application. The company behind the software had already issued a patch to fix the problem, but many users hadn't bothered to install it. The result? Hackers gleefully broke into thousands of systems, causing data breaches galore and racking up millions in damages.

It was the cybersecurity equivalent of locking your front door but leaving the back one wide open. This sobering example illustrates the peril of skipping updates. Each delay is an open invitation to cyber-criminals, who are only too happy to RSVP.

Making Updates Easy: Automate Where Possible

Here's the good news: you don't have to be glued to your screen, frantically checking for updates every day. Many systems and applications have embraced the magic of automation, offering automatic update features that do the heavy lifting for you. Enabling auto-updates ensures that critical patches are installed promptly, keeping your defenses strong while you sip your coffee or catch up on emails.

Here's how to make it happen:

- **Operating Systems**: Both Windows and macOS make automatic updates a breeze. You'll find the option in your system preferences or control panel—turn it on, and let the system do its thing.

- **Apps and Software**: Many applications have an "auto-update" toggle in their settings. Flip that switch, especially for programs that handle sensitive data or are heavily used by your team.

- **Antivirus Software**: This is a biggie. Antivirus programs need regular updates to stay ahead of the latest threats. Make sure yours is set to update automatically so it's always primed for battle.

With auto-updates enabled, you've taken a small but mighty step toward digital security. It's cybersecurity on autopilot, and who doesn't love that?

Creating an Update Schedule

Not every piece of software has the luxury of an auto-update feature, which is where a bit of old-fashioned routine comes in handy. Setting a regular update schedule—say, once a month—can make a world of difference.

Mark it on your calendar, set a reminder on your phone, or tie it to an existing routine (like checking the fire alarms or reconciling accounts). Take a few minutes each month to see if any updates are available. For businesses with multiple systems or team members, a shared checklist can be a lifesaver, ensuring that no program gets overlooked.

Think of this as the digital equivalent of brushing your teeth or changing your car's oil: routine maintenance that keeps everything running smoothly and prevents bigger problems down the road.

The Big Payoff

In cybersecurity, little habits like these add up to big results. Staying diligent about updates and patches makes your business a far less appealing target for hackers. It's a small effort with an outsized payoff: a safer, more resilient operation that's ready for whatever cyber threats come its way.

So, the next time that update notification pops up, don't hit "Remind Me Later." Take a moment to appreciate that it's not just a tedious chore—it's a powerful tool in your arsenal. And when you click "Update Now," you can smile, knowing you've just outsmarted a hacker or two. Bravo, you!

Backing Up Your Data: A Safety Net for When Things Go Wrong

Imagine the horror of losing your wallet—your ID, credit cards, even that receipt you swore you'd need someday. Stressful, right? Now, imagine that same gut-wrenching moment, but instead of your wallet, it's your business's data. Financial records, customer information, years of hard work—poof! Gone, thanks to a hardware failure, cyber attack, or someone accidentally clicking "delete."

This is where backups come to the rescue. Think of them as your safety net, always ready to catch you when life decides to toss a curveball. For small businesses, having a reliable backup system isn't just a good idea—it's essential. And the best part? It's easier to set up than you might think.

Why Backups are Essential for Small Businesses

Let's think of backups as your business's safety deposit box. You wouldn't leave your grandmother's jewelry or that irreplaceable stack of baseball cards lying around the house, would you? No, you'd put them in a secure spot where they're safe from burglars and floods. Backups do the same for your data.

Data loss can strike from all directions: a rogue hacker, a sudden hardware crash, or even a particularly enthusiastic employee accidentally wiping out the company's most important spreadsheet. With backups in place, these "oh no" moments transform into minor inconveniences. Instead of panicking, you simply click a few buttons, restore your files, and carry on. It's the digital equivalent of knowing you've got an extra tire in the trunk during a flat.

Types of Backups and When to Use Them

Not all backups are created equal. Depending on your needs, you'll want to pick the right type—or better yet, a combination. Let's break it down:

- **Full Backups**: Picture a full backup as your "big picture" solution. It's a complete copy of every last file, taking up more storage and time but ensuring you've got a full archive to fall back on. These are ideal for weekly or monthly snapshots when you want a comprehensive safety net. Think of it as your business's family portrait—everything captured in one go.

- **Incremental and Differential Backups**: These are the efficient little siblings of the full backup. Incremental backups save only the changes made since the last backup, making them faster and easier on storage space. Differential backups save everything changed since the last full backup, striking a balance between speed and comprehensiveness. These types are perfect for daily backups—they're like adding Post-it notes to your family portrait with updates about who's grown a few inches.

- **Cloud Backups vs. Physical Backups**: Now, where to store all this precious data? You've got two main options:

  - **Cloud Backups**: Services like Google Drive, Dropbox, or specialized backup providers let you save data in the digital ether. These backups are accessible from anywhere and protect against physical disasters like floods or fires.

  - **Physical Backups**: External hard drives and USB drives let you keep your data close at hand, without relying on an internet connection. They're not as trendy as cloud solutions, but they're reliable in their own way.

The golden rule? Use both. A mix of cloud and physical backups ensures you've got a backup for your backup. If one fails, the other is there to save the day.

By setting up a solid backup system, you're not just protecting your business—you're giving yourself peace of mind. Whether it's a sneaky hacker, a spilled cup of coffee, or sheer human error, backups ensure that no disaster is ever truly the end of the story. And really, isn't that worth the small effort it takes to get them in place?

How Often Should You Back Up Your Data?

Let's talk frequency. How often you should back up your data depends largely on how vital it is to your daily operations. Think of it as deciding how frequently to water your plants. Some thrive on daily attention, while others are fine with the occasional drink. Here's a simple guide to keep your digital greenery alive and well:

- **Critical Data** (like client information, financial records): These are your orchids—the delicate, high-maintenance stars of your operation. Aim for daily backups here. Sure, losing a day's work is frustrating, but it's infinitely better than losing a week's worth of critical data.

- **Less Critical Data** (internal communications, project files): These are your succulents—low-maintenance but still deserving of some care. Weekly backups usually do the trick for files that don't change every day but are still important to the business.

- **Non-Essential Data** (archived materials, historical data): These are your cacti. They're sturdy and can go longer without attention. A monthly backup will suffice to keep these files safe without hogging all your storage space.

Of course, you can tweak this schedule based on your specific needs, but sticking to this general rhythm keeps your data secure without overwhelming your system—or your sanity.

How to Test Your Backups

Now, let's imagine a nightmare scenario: you've been faithfully backing up your data, feeling wonderfully prepared, only to discover when disaster strikes that your backups don't actually work. It's the cybersecurity equivalent of realizing the fire extinguisher you bought is filled with glitter instead of foam.

This is why testing your backups is non-negotiable. Every so often, pick a few files at random, restore them from your backup, and make sure they open properly. It's a small, simple step that ensures your backup process is doing its job. Think of it like testing a smoke alarm—one quick check gives you the confidence that everything is in working order.

Wrapping It Up

Regular backups might not win you a standing ovation, but they are undeniably one of the most effective ways to safeguard your business from data loss. By understanding the different types of backups, following the golden 3-2-1 rule (three copies, two different storage types, one offsite), and testing your backups periodically, you're giving yourself the digital equivalent of a safety net.

And let's face it—data loss isn't a matter of *if* but *when*. So, when that fateful day comes, you'll be sitting pretty, knowing you've got a rock-solid plan in place. And that, dear reader, is a victory worth celebrating. Bravo, you!

A Cautionary Tale About Outdated Backup Practices

Let's start with a friendly but firm piece of advice: if you're still using old-fashioned backup tapes, it's time to give them a proper send-off into retirement. Tape backups are notorious for being unreliable, prone to degradation, and downright cranky when you need them most. Relying on them in today's digital age is like expecting a cassette tape to sound pristine after baking in your car's glove compartment for a decade. They might work for a while, but when they fail, they fail spectacularly—and usually at the worst possible moment.

Now, even if you've upgraded to more modern methods, onsite backups alone aren't enough. If you're storing external drives or other backup devices at your business location, you're still at risk. Fires, floods, theft, or even simple hardware failure can take out everything in one fell swoop. To avoid this, always keep multiple sets of backups, with at least one stored offsite in a secure location. A safety deposit box, a trusted employee's home, or a reliable cloud service can all serve as your digital lifeboats.

A Real-Life Lesson in Backup Woes

Here's a story to drive the point home. Years ago, I ran an IT administrative training class for IT Professionals. I had a client, a church administrator, that wanted to attend one of these classes in hopes of learning to do some of easier, frequent tasks, to save money on outside help. I covered all the things that IT Server Administrators have to know in order to service/maintain servers and networks, including backup best practices, testing backups, and checking logs to ensure everything was running smoothly. He seemed confident that he could handle that sort of thing so he took that responsibility on himself (delegating some of the work to the church secretary.

The church secretary had been faithfully (no pun intended) following a solid routine, creating three complete sets of backups. One stayed at the church, another went to her house, and the third was stored securely in a safety deposit box. This system had been in place for over a year, and everyone felt confident they were prepared for the unexpected.

Then, one morning, disaster struck. The church's server wouldn't boot up. A hard drive had failed. No problem, right? They had backups. Or so they thought.

We asked for the most recent backup, only to discover that the data was over six months old. Alarm bells rang. We checked the other two sets of backups, but all were equally outdated. What had gone wrong?

It turned out the backups had been failing for six months, but no one had noticed because the logs hadn't been checked, and the backups hadn't been tested. Despite her diligent routine, the secretary had no idea the system had stopped working. The church administrator sheepishly admitted he had never followed through on testing the backups or verifying the logs—a task he had learned about in my class but had let slip.

To make matters worse, the church secretary wasn't exactly forgiving when we explained the situation. She bluntly told the administrator she was ready to strangle him for not checking the backups. She then spent countless hours manually reentering six months of giving records into the system—time and effort that could have been saved with a functional backup plan.

The Takeaway

This cautionary tale is a reminder that having a backup system isn't enough. You need to actively check it, test it, and ensure it's working as intended. And for goodness' sake, don't rely solely on onsite backups or outdated methods like tape drives. Modern solutions, combined with diligent maintenance and testing, are the keys to avoiding the kind of nightmare this church experienced.

So, take a moment to review your backup strategy. Test those backups, check those logs, and ensure your data is safe no matter what. If you are not willing/able to diligently do the work, ask your IT provider if they are. Your future self—and your team—will thank you.

Putting It All Together: Your Cyber Hygiene Checklist

As we wrap up this chapter, let's pull everything together into a simple, actionable checklist. Good cyber hygiene doesn't need to be complicated, but it does require commitment. These essential habits will keep your digital environment as clean as a whistle and ready to withstand common cyber threats.

The Essential Cyber Hygiene Habits

Before we ride off into the digital sunset, let's take a moment to revisit the key practices we've covered—those fundamental bits of cyber wisdom that every business, no matter how big or small, should adopt. Think of these as the hygiene habits of your digital life, like flossing for your files or handwashing for your hardware. They're simple, effective, and utterly essential:

- **Strong Passwords**: Creating complex, unique passwords for each account makes it harder for hackers to get in.

- **Multi-Factor Authentication (MFA)**: Adding a second layer of protection for key accounts ensures that even if someone cracks a password, they can't get far.

- **Regular Updates**: Updating software, apps, and systems keeps vulnerabilities patched and minimizes the risk of breaches.

- **Data Backups**: Regular backups ensure you have a safety net if things go wrong, whether it's a cyber attack, system failure, or accidental deletion.

These practices are your cybersecurity ABCs—the foundational steps every business should follow to protect itself in this wild, ever-changing digital world. They're not flashy or complicated, but much like locking your doors at night, they're the simple, sensible measures that make a world of difference.

Sample Cyber Hygiene Routine: Consistency is Key

To keep your cyber hygiene in tip-top shape, think of it as a daily habit—like brushing your teeth, but for your digital assets. It doesn't have to be overwhelming or time-consuming. A little consistency goes a long way in keeping your business secure. Here's a straightforward routine to help you stay on top of things without losing your mind:

- **Weekly**: Check that backups are running smoothly and that any pending updates are installed. It only takes a few minutes, but it keeps your defenses strong.

- **Quarterly**: Update passwords on sensitive accounts and review your MFA settings to ensure they're active and working properly.

- **Annually**: Conduct a full review of your cyber hygiene practices. This is your chance to see what's working, what needs improvement, and to make sure all key accounts and systems are secure.

This routine takes the mountain of cybersecurity and breaks it down into manageable, bite-sized tasks. It's designed to keep you secure without requiring a massive time investment. Think of it as the equivalent of tidying up your desk at the end of the day—small efforts that prevent big messes later on.

Using Tools to Simplify Cyber Hygiene

Why slog through it all manually when there are clever tools to take the heavy lifting off your shoulders? Think of these as your digital sous-chefs, quietly handling the tedious bits while you focus on running your business like the master chef you are. Here are a few essentials:

- **Password Managers**: These marvels securely store your complex passwords, so you don't have to remember (or worse, write down) each one. It's like having a little black book, but for your passwords—and infinitely safer.

- **Update Managers**: Automate software updates wherever possible. No more remembering to click "Install now" every time an alert pops up. Let the tools handle it, and enjoy knowing your system is always up to date.

- **Cloud Storage for Backups**: Cloud solutions not only keep your data safely tucked away offsite but often run backups automatically. It's the peace of mind equivalent of a security blanket, only smarter.

These tools streamline the process, letting you focus on your business while they handle the background grunt work.

Building a Habit of Good Cyber Hygiene

Cyber hygiene isn't a one-and-done affair; it's a habit. By weaving these practices into your daily business operations, like payroll or inventory checks, you're setting up for long-term success.

Encourage your team to get on board, too. Building a culture of cyber hygiene across your organization creates a united front against threats. When everyone is on the same page, even the smallest actions—updating passwords, flagging suspicious emails—contribute to a safer, more resilient business environment.

Conclusion: Keep it Clean, Keep it Safe

As we close the chapter on cyber hygiene, let's tip our hats to the power of small, consistent actions. This isn't just about ticking off boxes; it's an investment in your business's future stability. Think of it as regular maintenance for your digital assets—keeping them secure, protecting your reputation, and saving you from the dreaded trifecta of time, money, and headaches.

The Long-Term Benefits of Cyber Hygiene

By committing to these foundational practices, you're setting your business up for resilience. Cyber hygiene isn't just about dodging disasters—it's about building a proactive defense. Each strong password, every software update, every backup—they're the building blocks of a secure, thriving operation. Over time, these small efforts protect your bottom line and help you sidestep the costly fallout of data breaches and cyber incidents.

Empowerment Through Small, Consistent Actions

Here's the good news: you don't need to be a tech wizard to make this work. Cyber hygiene is all about small steps done regularly. Think of each habit as adding another layer to your defenses—stronger passwords, automatic updates, routine backups. Together, these layers create a fortress around your business that help to keep the bad guys out and your peace of mind firmly intact.

So keep it clean, keep it safe, and know that you're actively safeguarding everything you've worked so hard to build.

Tease the Next Chapter

In the next chapter, we're shifting focus to one of your most valuable—and potentially vulnerable—assets: your team. We'll dive into the art of training employees and cultivating a cybersecurity-savvy culture in your business. By empowering your team, you'll build a collective defense stronger than any individual effort. So let's keep moving forward, one simple step at a time, toward a secure and thriving business environment.

TRAINING YOUR TEAM

CYBERSECURITY IS A GROUP EFFORT

Introduction to Team Cybersecurity

P icture this: your business is a ship—sturdy, dependable, and built to weather the unpredictable waters of the digital world. You've patched up the hull with strong passwords, reinforced the sails with multi-factor authentication, and calibrated the compass with regular software updates. A fine vessel, indeed. But here's the rub: no matter how seaworthy your ship is, it's only as reliable as the crew steering it. And that's where your team enters the picture.

Cybersecurity isn't just about firewalls, clever software patches, or the IT department's Herculean efforts in the background. It's about people—your people. Every email your employees open, every link they click, and every attachment they download can either fortify your ship or leave it vulnerable to a digital cannonball. Hackers know this, which is why they often target individuals rather than systems.

This chapter is about enlisting your team as your cybersecurity crew. Together, we'll dive into why cybersecurity is everyone's responsibility—not just the IT department's—and how to transform your employees into a formidable first line of defense. You'll learn how to arm your crew with the knowledge and tools they need to recognize threats, respond effectively, and make smarter choices online.

The good news? Cybersecurity doesn't have to feel like navigating through stormy seas. With the right training and a dash of teamwork, your employees can become your strongest asset instead of your weakest link. We're going to map out a culture where everyone, from the person greeting visitors at the front desk to the CEO making big decisions, plays a role in keeping your business secure.

So hoist the sails, gather the crew, and let's chart a course toward a safer, more secure operation. Because when it comes to cybersecurity, the stronger the crew, the safer the ship.

Why Cybersecurity is a Team Sport

Picture cybersecurity as a high-stakes game of soccer. Every player on the field has a role to play, and if everyone does their part, the team stands a fighting chance. The IT department? They're your goalkeeper, standing ready to block any threats that come too close to your goal. Managers and leaders? They're the defenders, organizing the team and keeping attackers at bay. The employees? They're your midfielders, the vital link between defense and offense, keeping the ball moving and the game in control. And your client-facing team? They're the strikers, engaging with the outside world, scoring goals, and driving your business forward.

Now imagine this: it's a nail-biter of a match. Your team is holding its own against a fierce opponent, everyone working together to protect the goal and advance the ball. But then, one player gets distracted. Maybe they didn't fully grasp the game plan or simply let their attention wander. They make an ill-timed pass, leaving the ball open for the opposing team to intercept. In the blink of an eye, the tide turns, and the opposition scores.

In the world of cybersecurity, that poorly timed pass might be an employee clicking on a suspicious link, reusing a weak password, or falling for a cleverly disguised phishing email. And no matter how brilliant your goalkeeper (the IT department) is, one misstep can leave your team wide open for attack.

The takeaway here? Cybersecurity isn't a solo act or even a duet. It's a full-blown ensemble performance. It's not just the IT team's responsibility, nor is it something the tech-savvy folks in the corner handle while everyone else carries on obliviously. Every single person, from the receptionist welcoming guests to the CEO making strategic decisions, has a crucial role to play. And hackers? They know this all too well. They're not always trying to outwit the goalkeeper—they're watching for that one distracted midfielder or inexperienced defender to drop the ball.

But here's the silver lining: teamwork wins games, and cybersecurity is no exception. When everyone on your team understands their role, has the right training, and works together, the organization becomes a force to be reckoned with. Your employees don't need to be cybersecurity wizards—they just need to be equipped with the tools, awareness, and confidence to recognize threats and respond smartly.

So, as you think about your organization's approach to cybersecurity, picture it as your game plan. It's not about relying on a lone star player to carry the team. It's about cultivating a united, informed, and well-practiced crew that can tackle any challenge. Because when every player steps up, the team becomes stronger, more resilient, and nearly impossible to defeat. And in this game, that's how you win.

Think of it Like This: Your Team as a Fortress

Imagine your business as a grand fortress. Every person—whether they're manning the front desk, handling day-to-day operations, or steering the company from the top—is a stone in the wall that shields your organization from outside threats. But as anyone who's watched a castle siege movie knows, a fortress is only as strong as its weakest point. A single crack—a loose stone, if you will—can bring the entire structure crumbling down. In the realm of cybersecurity, those cracks often show up when employees aren't prepared to recognize or respond to threats. And that's where training comes in—it's like reinforcing those stone walls with industrial-grade steel beams.

Here's the thing: hackers thrive on human error. It's not your high-tech firewalls or anti-virus software they're banking on; it's that someone on your team will make a mistake. Maybe it's clicking on a convincing phishing email, reusing a password for the umpteenth time, or trusting an attachment that seems just a little off. These seemingly minor lapses are often all it takes for a cybercriminal to slither past your defenses.

But imagine the alternative. What if your team could recognize those threats before they even gained a foothold? A phishing email? Spotted, flagged, and sent packing. A weak password? Swapped out for something so complex even the NSA would be impressed. That dodgy attachment? Left unopened, where it can do no harm. With the right training, your employees transform from potential liabilities into your strongest line of defense.

And let's be clear: training isn't about giving people a rulebook and hoping they'll follow it. It's about empowerment. When your team understands the *why* behind the rules, when they see themselves as active participants in your business's cybersecurity strategy, something magical happens. They start to care. They develop the confidence to make smarter decisions, the awareness to spot red flags, and the initiative to take an extra moment to question that suspicious email or set up multi-factor authentication.

Think of it like this: while technology can protect you up to a point, it's the human element that makes or breaks your defenses. Your firewalls and software are the drawbridge and moat of your fortress, but your people are the guards on the wall. When every single one of them is equipped, informed, and ready to act, your fortress becomes nearly impenetrable.

By building a culture of security—one where everyone feels invested in keeping the business safe—you're not just defending against today's threats. You're creating a workplace where vigilance is second nature, where every team member is part of a united front against cyber risks. And in this world of ever-evolving threats, that kind of strength is invaluable.

Cybersecurity: More than Just an IT Concern

Cybersecurity often gets shunted off to the IT department, as if it's some kind of arcane art performed in rooms full of glowing screens and ominous cables. But here's the thing: it's not just their domain. It's everyone's concern. Picture your business as a grand castle—not the fairy tale kind with dragons and wizards, but a sleek, modern fortress filled with priceless treasures: your data, your reputation, and your clients' trust. Sure, the IT team are the master masons, building firewalls, installing antivirus software, and constructing secure networks. But what happens if someone accidentally leaves the drawbridge down?

Cybersecurity isn't just about what's happening behind the scenes. Every person in your organization—from the new hire who just learned where the coffee machine is to the CEO—has a part to play. It's the receptionist who might open a convincing phishing email. It's the sales manager who reuses the same password for work and their go-to streaming service. It's even the big boss who forgets to double-check before approving a suspicious wire transfer request. The point is, cybersecurity is a collective effort. Every action and decision either reinforces your defenses or creates a potential opening for cyber troublemakers.

When everyone in your organization understands their role in cybersecurity, something transformative happens. It stops being that mysterious "IT thing" and becomes a shared mission. Instead of a solitary watchman on the tower, you have an entire team, alert and ready. This culture of vigilance turns your business into a fortress where every person is a guardian, playing their part to keep things safe.

So, whether your desk is in the lobby or the corner office, remember this: you're not just a bystander in the realm of cybersecurity. You're a key player. Your choices, awareness, and actions matter more than you might think. Cybersecurity isn't just about firewalls and fancy software—it's about people, teamwork, and shared responsibility. When everyone steps up, the castle doesn't just survive—it thrives. And that's the kind of fortress worth building.

Building a Cyber-Aware Culture: Beyond the Checkbox

Creating a cyber-aware culture isn't about trudging through another tedious training session or forwarding the obligatory IT email about changing your password—again. No, it's about something much bigger. It's about weaving cybersecurity into the everyday rhythm of your organization, turning it into a natural reflex, like locking the front door when you head out or giving your seatbelt a tug before hitting the road. It's about cultivating habits that feel less like a chore and more like common sense.

A cyber-aware culture doesn't stop at a one-and-done training module. It lives in the small, consistent actions your team takes every day: pausing before clicking a link, questioning an unusual request, or flagging something that feels off. It's about giving people the confidence to ask, "Is this safe?" rather than assuming everything is fine.

Here's the kicker: this kind of culture isn't a "nice-to-have" bonus, like free snacks in the breakroom. It's absolutely essential for any business that wants to stay competitive and resilient in today's digital minefield. Cyber threats aren't going away, and technology alone isn't enough to keep them at bay. Your greatest defense is a team that knows how to spot risks, understands their role in mitigating them, and takes cybersecurity as seriously as they take their coffee breaks.

By embedding cybersecurity into your organization's DNA, you're not just protecting your data—you're safeguarding your reputation, your operations, and your future. A cyber-aware culture isn't just smart; it's the secret sauce that turns a good business into a great, secure one.

What a Cyber-Aware Culture Looks Like

Picture this: a workplace where every single employee—from the wide-eyed intern on their first day to the seasoned CEO—is on the same page about cybersecurity. Not just in a vague, "Oh yeah, I've heard of phishing" kind of way, but in a tangible, practical sense. They know what a suspicious email looks like. They pause before clicking links that seem even slightly off. And they don't hesitate to ask questions or report something that doesn't sit right. It's a place where people see cybersecurity not as a nuisance but as a collective responsibility—a team sport where everyone plays a role.

In this kind of culture, vigilance is second nature. It's an environment where no one is intimidated by tech talk or afraid of looking silly for double-checking a request. Instead, employees feel empowered and confident in their role as protectors of the company's digital assets. The goal here isn't to transform everyone into IT wizards or cybersecurity masterminds—though hats off to anyone who wants to try—but to ensure that everyone understands the basics. Simple actions, like setting a strong password or flagging a questionable email, add up to a formidable defense.

The beauty of a cyber-aware culture is that it's not just about ticking boxes or following rules. It's a shared mindset, one that turns good security habits into effortless, everyday behaviors. When cybersecurity becomes second nature, it stops being a chore and starts being a point of pride. The strongest defense, after all, doesn't just come from firewalls and fancy software—it comes from the people who use them wisely, every single day. And that's the kind of team you want protecting your business.

Leadership: The Linchpin of a Cyber-Aware Culture

When it comes to fostering a culture of cybersecurity, leadership isn't just important—it's critical. Why? Because employees take their cues from the top. If the people in charge treat cybersecurity like a box to tick or, worse, a nuisance, it sends a clear message: "This doesn't really matter." On the flip side, when leaders prioritize cybersecurity, it becomes part of the company's DNA—a shared responsibility woven into every aspect of the business.

Imagine a CEO who insists on using "Password123" for every account and casually clicks on suspicious links without a second thought. What does that signal to the rest of the team? It's like a captain of a ship dismissing the need for lifeboats. Conversely, when leadership models good cybersecurity habits—using strong, unique passwords, enabling multi-factor authentication, and pausing to scrutinize unexpected emails—it sets a standard. It says, "This is how we protect our business, our customers, and each other."

But modeling behavior is just the start. Great leaders take the time to explain *why* cybersecurity matters. Saying, "Be careful with emails," is one thing. Explaining that a single careless click could lead to a data breach, expose sensitive customer information, and cost the company thousands (or worse) drives the point home. When leaders connect the dots between individual actions and company-wide consequences, they transform cybersecurity from an abstract IT problem into a shared priority that everyone can rally around.

Leadership also means actively engaging in cybersecurity efforts, not delegating it to the IT department and hoping for the best. This could be as simple as attending training sessions alongside employees, initiating open discussions about potential threats, or even sharing stories of near-misses and lessons learned. When leaders roll up their sleeves and take an active role, it fosters a sense of accountability and teamwork. It shows that cybersecurity isn't just "their job"; it's *our* job.

Setting the tone for cyber awareness is about more than policies or procedures—it's about leading by example. Leaders who take cybersecurity seriously inspire their teams to do the same. They create an environment where employees feel empowered to stay vigilant, report concerns, and embrace their role in protecting the business.

Here's the bottom line: A strong cybersecurity culture doesn't start in the IT department. It starts in the corner office, in the boardroom, and with every manager who champions the cause. When leadership is all in, the rest of the organization follows suit. And that's how you build not just a secure business, but a resilient one.

Creating a Culture of Cybersecurity Curiosity

Cybersecurity isn't exactly the sort of thing people chat about over coffee in the break room. For many employees, the topic feels about as inviting as a pop quiz on quantum physics. There's often a worry—perhaps unspoken—that asking a "simple" question about a sketchy email or a muddled password might make them look foolish or unprepared. And that's precisely why fostering a culture of open communication around cybersecurity is so vital.

Imagine a workplace where employees feel entirely at ease asking questions like, "Is this email safe to open?" or, "What's the best way to handle this file with all these sensitive details?" It's not about expecting everyone to know everything—it's about encouraging curiosity and creating a supportive space where no one feels embarrassed for speaking up. After all, even the savviest tech wizards started somewhere, and the only silly question in cybersecurity is the one you didn't ask.

Here's the thing: when people feel comfortable asking questions, they're far more likely to engage with cybersecurity practices. They'll pause before clicking, double-check before sharing, and generally approach their digital interactions with an extra layer of care. That small moment of hesitation—a quick "Hey, does this look legit?"—could be the difference between business as usual and a full-blown crisis.

Encouraging open communication doesn't just make employees feel supported; it drastically reduces the likelihood of mistakes. And let's face it, the stakes are high. Hackers thrive on uncertainty and hesitation, counting on people to click first and think later. By fostering an environment where questions are welcomed, you're essentially building an internal safety net—a collective effort that catches potential issues before they escalate.

So, let's retire the idea that cybersecurity is a solo sport or that asking questions signals weakness. It's quite the opposite. A curious, communicative team isn't just better informed; it's better protected. And in a world where digital threats are constantly evolving, that's a win for everyone. After all, isn't it far better to ask one awkward question than to scramble through the aftermath of a breach? I think we both know the answer.

The Art of Effortless Cybersecurity

The beauty of good cybersecurity practices lies in their simplicity—they don't need to upend your day or slow down your work. In fact, they work best when they're seamlessly woven into daily routines, becoming as automatic as brushing your teeth or grabbing your keys before leaving the house. Small, thoughtful actions—like locking your computer when you step away, pausing to verify an email's sender before clicking on a link, or using secure file-sharing platforms—can quietly create a fortress of security over time.

Think of these practices as the digital equivalent of "look both ways before crossing the street." They don't require much effort, but they offer immense protection. And just like crossing the street safely, they quickly become second nature when gently reinforced. A quick reminder in a team meeting or a strategically placed poster in the breakroom can subtly nudge employees to keep these habits top of mind without ever feeling intrusive.

Over time, these small steps start to feel less like "extra work" and more like an effortless part of the daily rhythm. It's not about being hypervigilant or paranoid—it's about being aware and intentional. When these habits are baked into how employees operate, cyberse- curity stops feeling like a chore and starts becoming just another way the team works smartly and securely. It's proof that protecting your business doesn't always require heroic efforts—sometimes, it's the little things, done consistently, that make the biggest difference.

Making Cybersecurity Celebrations a Team Tradition

Cybersecurity might not sparkle like the launch of a new product or the clinching of a big sale, but that doesn't mean it can't be re- warding—or even fun. Recognizing and celebrating employees who practice good cybersecurity habits is one of the best ways to reinforce positive behavior and show your team that their efforts matter.

Imagine this: Someone on your team catches a phishing at- tempt before it can wreak havoc. Why not give them a well-deserved shout-out at the next team meeting? Or perhaps a colleague takes the initiative to create unique, complex passwords for their accounts without anyone nudging them. Maybe that's worthy of a small gift card, or even just a genuine pat on the back and a few kind words. These little acknowledgments might seem small, but they go a long way in making cybersecurity feel less like a chore and more like an achievement.

When employees see their efforts recognized, it creates a ripple effect. The team starts to notice that cybersecurity isn't just about avoiding mistakes; it's about taking proactive steps that benefit everyone. And those moments of recognition encourage others to step up, fostering a culture where good habits become contagious.

Celebrating wins, no matter how small, transforms cybersecurity from a looming responsibility into an approachable and engaging part of the workplace. It's not just about strengthening your defenses—though that's a fantastic bonus. It's about building a team that feels valued, confident, and united in protecting the organization. And that's a victory worth celebrating every single time.

Reimagining Cybersecurity Training: Yes, It Can Be Fun

When the phrase *"cybersecurity training"* comes up, you can almost hear the collective groans and see the mental images of endless Power-Point slides, delivered in the kind of monotone that could lull even the most alert coffee drinker into a nap. But here's the thing—it doesn't have to be that way. Cybersecurity training can be interactive, engaging, and, dare I say it, fun. Yes, *fun*. And before you roll your eyes, hear me out: when learning is enjoyable, it sticks.

Think of it like gamifying safety drills. What if instead of sitting through a dry lecture about phishing emails, employees competed in a "spot the scam" challenge? Or participated in a simulated ransomware attack where their quick thinking could save the day (or at least a fictional company)? When training feels less like a chore and more like a team-building adventure, people are not only more willing to engage—they actually retain what they've learned.

The goal isn't to sugarcoat serious issues but to create an environment where employees feel empowered to learn and apply good cybersecurity habits. Interactive workshops, quick quizzes with prizes, or even cybersecurity-themed escape rooms (yes, they exist!) can transform training from a checkbox exercise into a valuable experience. Because when people are having fun, they're paying attention—and that's half the battle won.

So, ditch the droning presentations and think creatively. Cybersecurity training doesn't have to be a slog; it can be something your team genuinely looks forward to. Who knows? With the right approach, your next training session might just be the highlight of the workweek.

Rewriting Cybersecurity Training: From Chore to Cheer

Let's face it—nobody eagerly anticipates mandatory, dry training sessions. It's human nature. Toss in the word *cybersecurity*, and you might even see a few dramatic sighs or exaggerated eye rolls. But here's the reality: cybersecurity is far too important to let it drift into the *"just another thing"* category. The good news? It doesn't have to stay there.

Inject a little creativity, humor, and interactivity into the mix, and you can transform what might have felt like a slog into something employees genuinely look forward to. Imagine turning the typical "spot the phishing email" task into a game, complete with prizes for the sharpest eyes. Or holding a team competition to see who can create the most hilariously complex (yet secure) password. Add a bit of fun, and suddenly, you're not just teaching your team; you're engaging them.

It's not just about keeping attention—it's about making the material stick. No one's going to forget the phishing drill where the winner walked away with bragging rights *and* a coffee gift card. When training feels like an experience rather than an obligation, employees are far more likely to absorb the information and put it into practice.

So, let's wave goodbye to monotonous lectures and say hello to an approach that makes cybersecurity training memorable, effective, and yes—even enjoyable. Because when the stakes are this high, the method matters. And a little fun goes a long way.

Phishing Simulations: Your Cybersecurity Dress Rehearsal

Think of phishing simulations as the *dry run* for the digital world's trickiest con games. These exercises are like a dress rehearsal for real-life cyber threats, complete with all the drama but none of the real-world consequences. Here's the idea: you send out realistic (but totally fake) phishing emails to your employees, not to catch them off guard or humiliate anyone, but to give them a safe space to learn and sharpen their instincts.

When someone takes the bait, it's not the end of the world—it's the beginning of a valuable lesson. They get a chance to see where they went astray, maybe even laugh it off, and gain practical insights to avoid falling for the real thing. And those who spot the scam? They walk away with a well-earned boost in confidence and the quiet satisfaction of knowing their vigilance paid off.

The beauty of simulations is how they fine-tune that all-important sixth sense for phishing. Employees learn to spot the giveaways—like email addresses that don't quite add up, an oddly frantic tone, or that too-good-to-be-true link. Over time, what starts as cautious hesitation evolves into confident, automatic caution.

Phishing simulations aren't about assigning blame; they're about building awareness. The more you practice, the better your team gets at sidestepping those digital traps—and that's a win for everyone.

Phishing Simulations on a Shoestring Budget

Good news: you don't need a blockbuster budget or cutting-edge tech to pull off an effective phishing simulation. With a dash of creativity and a sprinkle of strategy, you can run a simulation that's as enlightening as it is economical. Start small and straightforward—create a fake but convincing phishing email using classic tricks like, "Your account has been suspended—click here to reset your password!" It's a timeless favorite in the hacker's handbook for a reason.

Once your email masterpiece is ready, send it to your team and sit back to see who takes the bait. But here's the crucial part—this isn't about shaming anyone or wagging fingers. When someone clicks the link, follow up with a friendly, "Gotcha!" email that explains what went wrong, why it worked, and how they can sidestep similar traps in the future.

The goal here is to turn mistakes into meaningful lessons. Treat the whole experience as an opportunity to educate, not reprimand. This is your chance to build awareness and arm your team with practical skills for spotting phishing attempts in the wild.

And don't forget to celebrate progress! Track how your team improves over time and acknowledge those who nail the art of spotting scams. A bit of recognition goes a long way in reinforcing good habits and turning cybersecurity into something everyone feels confident about. Remember, this is a learning journey—not an exam—and every step forward makes your organization a little more secure.

Cybersecurity: Let the Games Begin

Who doesn't enjoy a bit of friendly competition? Adding gamification to your cybersecurity training can turn it from a collective groan into something people actually look forward to. Picture this: a trivia showdown where employees race to answer cybersecurity questions, or a scavenger hunt where they comb through sample emails to spot phishing red flags. Toss in a prize or two—think coffee gift cards, an extra-long lunch break, or, best of all, workplace bragging rights—and you've got yourself an event that's as engaging as it is educational.

The beauty of games and quizzes is that they don't just entertain—they solidify learning. When employees are actively engaged, they retain information better, making them sharper and more confident when faced with real-world cyber threats. Plus, a little playful competition fosters teamwork and makes cybersecurity training feel less like homework and more like a team-building exercise.

So, get creative. Turn identifying phishing attempts into a points-based challenge, or award badges for mastering secure password creation. With a dash of imagination, cybersecurity can go from mundane to memorable. And when it's fun, it sticks.

Step Into Character: Role-Playing Your Way to Cybersecurity Smarts

Want to make cybersecurity training a little more... theatrical? Enter role-playing exercises—a surefire way to bring lessons to life. Imagine this: one employee plays the unsuspecting recipient of a suspicious email, while another channels their inner hacker, crafting a devious message. The rest of the team? They're the cybersecurity think tank, ready to analyze and advise.

The goal is to work through the recipient's next steps in real time. Should they verify the sender? Report the email to IT? Or simply hit delete and move on with their day? These exercises give employees a chance to practice real-world responses in a safe, no-pressure environment.

Role-playing doesn't just teach; it sticks. By physically stepping into the scenario, employees internalize the steps they need to take when the stakes are real. Plus, it's a collaborative experience that gets the whole team involved, reinforcing the idea that cybersecurity isn't just an individual responsibility—it's a team effort.

And let's face it, who doesn't love a little workplace drama when it's all in good fun? These exercises might just turn the drudgery of cybersecurity training into the highlight of the week.

Cybersecurity Training Made Easy: Online Resources at Your Fingertips

If the idea of organizing games or role-playing makes you break out in a sweat, fear not. The internet is brimming with user-friendly tools and resources to make cybersecurity training both practical and approachable. These platforms cater to all learning styles, offering everything from bite-sized videos to full-fledged courses, all designed to fit seamlessly into your team's busy schedules.

Imagine a quick five-minute refresher video on spotting phishing attempts—simple, engaging, and perfect for a coffee break. Or a short, interactive quiz that walks employees through the essentials of multi-factor authentication without inducing a single yawn. These resources deliver valuable lessons without the need for elaborate setups or marathon training sessions.

Many platforms even offer free content, making this an accessible option for businesses of any size. Employees can tackle modules at their own pace, whether it's a crash course on password security or an in-depth dive into email safety. It's training on their terms, but with all the benefits of structured learning.

The beauty of these tools is their flexibility. You can assign them as part of an onboarding process, use them for quarterly refreshers, or simply have them available for self-guided learning. With these resources, cybersecurity training doesn't have to be a hassle—it's just another way to empower your team to protect the business.

As a starter, here are five organizations that offer free cybersecurity training:

1. **Cybersecurity and Infrastructure Security Agency (CISA)**CISA, part of the U.S. Department of Homeland Security, provides free cybersecurity training resources. They offer self-paced courses, webinars, and tools focused on building cybersecurity awareness and skills.

2. **SANS Cyber Aces Online**SANS is one of the most respected names in cybersecurity training. Their Cyber Aces Online program is free and covers foundational topics like networking, operating systems, and security essentials.

3. **IBM SkillsBuild**IBM offers free courses through its Skills-Build platform, including a cybersecurity pathway that introduces learners to key concepts and tools. This resource is excellent for beginners and professionals looking to upskill.

4. **Fortinet NSE Training Institute**Fortinet provides free cybersecurity training through its Network Security Expert (NSE) Certification program. The NSE 1 and NSE 2 courses are free and cover cybersecurity awareness and foundational skills.

5. **OpenLearn by The Open University**OpenLearn offers free courses on various topics, including cybersecurity. Courses like "Introduction to Cyber Security" provide a solid grounding in essential cybersecurity principles and practices.

These resources are valuable for individuals and teams looking to enhance their cybersecurity knowledge without breaking the budget.

Cybersecurity Training: Learning Meets Engagement

Cybersecurity training doesn't have to feel like a trip to the dentist—necessary but unenjoyable. Instead, think of it as an opportunity to engage your team in a way that's memorable and, hopefully, fun. By weaving in simulations, games, and interactive scenarios, you're transforming what could be a dry lecture into a hands-on experience that sticks.

Picture this: a phishing simulation that feels like a detective game, or a trivia showdown where employees flex their cybersecurity knowledge for bragging rights and maybe a coffee gift card. These activities do more than entertain—they immerse your team in real-world situations, giving them the confidence to recognize and respond to threats when it matters most.

Interactive scenarios, like role-playing exercises, help employees step into the shoes of both attackers and defenders. They don't just learn the "what" of cybersecurity—they experience the "how" in a safe, low-pressure setting. This kind of active participation transforms training from a chore into something your team might actually look forward to.

The more engaged your employees are, the more effective your cybersecurity defenses become. After all, a team that's invested in learning is a team that's ready to defend your business from whatever digital threats come its way. And here's the best part: with a little creativity, you'll prove that cybersecurity training doesn't just protect your business—it can even be fun. Who knew?

Cybersecurity: The Art of Nipping Trouble in the Bud

Cybersecurity is a bit like spotting a tiny leak before it floods your kitchen. It's all about catching problems early, before they snowball into disasters that leave you mopping up for weeks. And one of the simplest, yet most effective tools in your arsenal is the power of reporting.

Encouraging your employees to report anything even slightly suspicious—a rogue email, an unexpected pop-up, or a fishy-looking link—can be the cybersecurity equivalent of shutting off the water supply before the sink overflows. It's proactive, it's practical, and it saves you from having to deal with a soggy mess (metaphorically speaking).

The beauty of reporting is that it doesn't require high-tech expertise or fancy software. It's about building a culture where people feel comfortable raising a hand and saying, "Hey, this doesn't look right." Because, let's face it, stopping an attack before it starts is infinitely better than scrambling to clean up after the fact. So, let's make reporting less of a "nice-to-have" and more of a reflex—like checking your keys before locking the door.

The Cybersecurity Fire Drill

Imagine this: an employee clicks on an email attachment that looks perfectly innocent—maybe an invoice from a "client" or a cheerful message promising free coffee. But, alas, it's not innocent. It's the digital equivalent of a Trojan horse, carrying malware straight into your system. Now, here's the twist: the quicker that employee raises the alarm, the faster your IT team can swoop in like digital firefighters. They can isolate the infected system, contain the threat, and prevent the malware from spreading like a bad rumor.

Early reporting isn't just a safety net—it's a lifeline. It transforms what could've been a catastrophic breach into a manageable hiccup. Think of it as catching a small kitchen fire before it consumes the curtains.

And let's take it a step further. Reporting is like noticing the faint smell of smoke in a building. If someone speaks up right away, there's a chance to find the source and douse the flames before they spread. But if everyone shrugs it off, assuming it's not a big deal, the whole place could end up as a smoldering ruin.

In cybersecurity, silence isn't golden; it's combustible. Encouraging employees to report quickly and confidently is one of the best ways to keep your business safe. After all, it's much easier to put out a spark than rebuild after a blaze.

Let's Be Honest: The Fear of Reporting

No one wants to be *that person*—you know, the one who raises the alarm over something harmless or sheepishly admits they might've clicked on a questionable link. It's human nature to want to avoid embarrassment or, worse, punishment. But here's the truth: staying silent and crossing your fingers is a gamble, and in the cybersecurity world, it's one you don't want to take.

The good news? Reporting isn't about blame—it's about safeguarding the entire organization. It's not a game of "Gotcha!" but rather a critical step in protecting your business from potential harm. A suspicious email? A strange pop-up? Even if it turns out to be nothing, reporting it is a sign that your team is paying attention. And that's exactly what you want—engaged, alert employees who take cybersecurity seriously.

Mistakes happen—we're human, after all—but ignoring them? That's when the real trouble begins. By reassuring your team that it's okay to speak up, you're fostering a culture of vigilance rather than fear. After all, it's far better to deal with a false alarm than to discover, too late, that silence allowed a minor issue to escalate into a full-blown crisis.

Fostering a Judgment-Free Zone: Encouraging Reporting

To create an environment where reporting thrives, your organization needs to be a no-blame zone—a place where employees feel safe admitting mistakes or asking questions without the dread of reprimand. Let's face it: nobody's perfect, and in the fast-paced world of emails, pop-ups, and countless daily clicks, errors are inevitable. But here's the thing: what happens after the mistake makes all the difference.

Leaders set the tone. If an employee comes forward and admits they've clicked on a suspicious link, the response shouldn't feel like a trip to the principal's office. Instead, it should go something like this: thank them for reporting the issue, address it promptly, and then turn it into a constructive learning moment for the team. "Hey, this could've happened to anyone. Let's use this as a chance to sharpen our phishing radar."

When employees know they'll be met with support instead of blame, they'll be far more likely to raise their hand when something seems off. And in the high-stakes game of cybersecurity, those few extra seconds of awareness and action can be the difference between a near miss and a full-blown crisis.

Remember, reporting isn't about pointing fingers; it's about keeping the organization secure. A culture of understanding doesn't just encourage employees to speak up—it builds a united front where everyone feels empowered to play their part in protecting the business. And that's a win for everyone.

Simplifying the Reporting Process: A Clear Path to Action

Creating a judgment-free culture is essential, but it's only half the battle. If employees are left scratching their heads about who to call or what to do when they spot something suspicious, even the best intentions can fizzle out. That's why having a crystal-clear, easy-to-follow reporting process is just as vital as fostering an open environment.

Think of it like a fire drill: everyone knows where to go, what to do, and who's in charge. The same should apply to reporting potential cybersecurity threats. Employees shouldn't have to play a guessing game when the stakes are high.

Start by establishing a designated point of contact. This could be a specific email address, a Slack channel, or even a dedicated team member who serves as the go-to for all things suspicious. Then, make the process ridiculously simple and impossible to forget. For example:

1. **Spot something suspicious?** Don't panic—stay calm.

2. **Report it to [designated contact method].** Whether it's an email or a quick message on Slack, make the pathway clear.

3. **Include as much detail as possible.** A screenshot, the exact time it happened, or even just a quick description can make all the difference.

The goal is to make reporting feel like second nature. When employees know exactly what to do, they're more likely to act quickly and decisively—turning potential threats into manageable moments rather than full-blown crises.

And don't forget to communicate the process often. Post it in common areas, pin it in team channels, or include it in onboarding materials. The easier and more familiar it is, the faster and more effective your team's response will be. Because when reporting becomes second nature, you're not just reacting to threats—you're staying one step ahead.

Turning "Near Misses" into Golden Opportunities

Not every potential security incident turns into a disaster—and that, my friends, is something to celebrate. These "near misses," like an employee hovering over a phishing link but deciding not to click, are more than close calls; they're invaluable teaching moments. Instead of viewing them as brushes with doom, treat them as opportunities to reinforce good habits and build a culture of awareness—without the shadow of fear or blame.

Imagine this scenario: an employee receives a dodgy email, senses something off, and reports it before clicking. That's a cybersecurity win, plain and simple. Take a moment to celebrate their vigilance! A quick acknowledgment in a team meeting or a congratulatory email can work wonders. It shows that their attention to detail matters and makes others more likely to follow suit.

For even greater impact, share these "near miss" stories with the rest of the team. Keep it anonymous if you need to, but lay out the details: what happened, how it was spotted, and why it mattered. It's a low-stakes way to reinforce the importance of staying alert, and it helps employees visualize what to watch out for.

Think of it as the workplace equivalent of sharing tales about narrowly avoided mishaps, like almost leaving your wallet behind on the train but grabbing it just in time. These moments stick with people, subtly shaping their behavior and making them more cautious in the future.

Near misses are the perfect balance of education and encouragement. They remind everyone to stay on their toes while fostering a sense of teamwork and shared responsibility. So, the next time an employee reports a suspicious email or sidesteps a potential security misstep, don't let the moment pass by. Celebrate it, share it, and turn it into a lesson for the entire team. After all, every near miss is a step closer to a safer, stronger organization.

Reporting: A Proactive Play for Cybersecurity Success

Let's dispel a common misconception: reporting isn't just something you do when things go wrong. It's not about waving a white flag after the fact; it's a proactive step—an essential move in your business's ongoing effort to stay secure. When employees are empowered to report suspicious activity, supported when they make mistakes, and confident that their actions make a difference, you're not just solving problems—you're preventing them.

Creating this kind of environment takes more than an occasional email reminder. It's about building a culture where everyone feels like they're part of the team. When employees know they won't be judged for speaking up, they're far more likely to report that strange email, odd pop-up, or unusual network activity. And when they do, those reports become invaluable tools for spotting potential threats before they can wreak havoc.

Think of reporting as one of the most valuable plays in the cybersecurity game. It's not flashy, like a game-winning goal or a last-minute save, but it's the steady, dependable teamwork that keeps your business on solid ground. Every time someone reports something suspicious, they're contributing to a collective defense—strengthening the walls of your digital fortress brick by brick.

The beauty of this proactive approach is that it turns cybersecurity into a shared responsibility. It's no longer something relegated to the IT team or buried in the fine print of an employee handbook. It's alive in the actions of every person in your organization.

So, the next time an employee spots and reports something unusual, take a moment to acknowledge the importance of their contribution. Let them know that their vigilance makes a difference—not just for their own work but for the entire organization. Because in the grand scheme of cybersecurity, reporting isn't just a safety measure. It's a cornerstone of a culture that's ready to take on any threat.

A Hard Truth About Cybersecurity: Lessons from the Frontlines

I wish I could tell you that everything I've shared so far—about phishing emails, ransomware, and the critical role your team plays in cybersecurity—was purely theoretical. That it was all just a collection of cautionary tales meant to spark your imagination and drive the point home. But, unfortunately, it's not. Every scenario I've described is rooted in real-world experiences. Over the years, we've worked with hundreds of companies, offering advice, laying out strategies, and, at times, issuing dire warnings about the importance of taking cybersecurity seriously.

And here's the thing: many listened. They saw the risks clearly, took proactive steps, and built the defenses they needed to protect their businesses. But not everyone did. Some shrugged it off, convinced that cybersecurity was something for "other companies" to worry about. "We're too small to be a target," they'd say, or, "We don't have anything hackers would want." These leaders chose to take the gamble—and I wish I could say they beat the odds.

They didn't.

For those who dismissed the warnings, the outcomes were often catastrophic. I've seen leaders lose their jobs after breaches exposed sensitive client data. I've witnessed businesses buckle under the weight of financial losses, unable to recover after a cyberattack drained their resources. Perhaps most heartbreaking of all, I've watched as communities and employees felt the fallout from decisions that could have been avoided with just a little foresight and action.

It's not just the immediate damage to a company's bottom line that makes cybersecurity failures so devastating—it's the ripple effects. Clients lose trust, employees face uncertainty, and years of hard work can crumble overnight. The leaders who brushed off the risks didn't just pay with their profits; they paid with their reputations, relationships, and sometimes, the very survival of their businesses.

So, here's my plea: don't make the same mistake. Cybersecurity isn't an abstract problem or a "nice to have" feature of your operations. It's a necessity—a fundamental part of protecting what you've built.

The risks are real, the stakes are high, and the consequences of inaction are severe.

This isn't about scaring you into submission. It's about giving you the tools, knowledge, and motivation to take action before it's too late. Because the truth is, you don't want to find yourself looking back, wishing you'd taken cybersecurity seriously when you still had the chance. The time to act is now, and the choice is yours. Let's make it the right one.

A Cybersecurity Policy: Your Business's Digital Playbook

A cybersecurity policy is like the rulebook for a game—except the stakes are much higher than a trophy or bragging rights. It's your business's official guide to navigating the digital world safely, spelling out the dos and don'ts that help protect sensitive data and fend off cyber threats. Think of it as the playbook that ensures everyone on your team knows how to handle their online interactions responsibly and securely.

Without a policy, you're essentially sending your team onto the field without a clue about the rules. It's chaotic at best and dangerous at worst. Imagine a defender scoring an own goal because they didn't know which direction to aim or a player wandering off the field mid-game. That's what happens when employees aren't clear about cybersecurity expectations—it leaves your business vulnerable to preventable mistakes.

A strong cybersecurity policy doesn't just set boundaries; it empowers your team. It provides clarity on how to handle everything from creating passwords to identifying phishing attempts, making cybersecurity less of a mystery and more of a shared responsibility.

And in today's digital landscape, having everyone on the same page isn't just a nice-to-have—it's essential. Because when everyone knows the rules, your business is better equipped to defend against whatever threats come its way.

Why Every Business Needs a Cybersecurity Policy

No matter if your business has two employees or two thousand, a cybersecurity policy is essential. It's the guidebook that keeps everyone—from the summer intern fetching coffee to the CEO making high-stakes decisions—on the same page. A solid policy not only provides clarity but also sets expectations, removing the guesswork from navigating the digital world.

Think of it as the definitive answer key for all those lingering questions: How should passwords be created and stored? What's acceptable to do on a company device (and what's definitely not)? And most importantly, what steps should someone take if they think something has gone awry?

By addressing these questions upfront, your cybersecurity policy becomes more than just a document. It's a shared agreement that reduces uncertainty, ensures consistency, and empowers your team to work confidently and securely in today's digital-first landscape. With a clear policy in place, you're not just setting expectations—you're building a stronger, safer foundation for your business.

Crafting a Cybersecurity Policy: The Essentials Every Business Needs

Even if your business is just dipping its toes into the waters of cybersecurity, your policy needs to cover some fundamental ground rules. Think of these as the non-negotiables—the digital equivalent of locking the office door when you leave for the night. Let's break it down:

Password Requirements

Weak passwords are practically an open invitation to hackers. Your policy should lay out clear guidelines for password length (hint: longer is always better), complexity (a mix of uppercase letters, lowercase letters, numbers, and symbols), and how often they should be updated. And let's be clear: unique passwords for different accounts aren't just a "nice-to-have" anymore—they're essential.

Data Handling Guidelines

Sensitive data—be it customer information, employee records, or financial details—deserves VIP treatment. Your policy should spell out how to handle it securely, from encryption for digital files to using secure cloud storage for backups. And no, jotting down sensitive information on sticky notes doesn't count as a secure method.

Acceptable Use of Company Devices

What's fair game on a company laptop or network? Your policy should spell this out in no uncertain terms. Make it clear that personal downloads, risky websites, and non-work activities on work devices are big no-nos. Think of it as setting the digital house rules.

Incident Reporting Procedures

When something seems off—a phishing email, an odd pop-up, or a system acting like it has a mind of its own—your team needs to know exactly what to do. Include simple, actionable steps for reporting potential issues and identify a go-to person or department. Speed matters in these situations, and a clear process makes all the difference.

Email and Internet Usage

Emails and unsecured browsing are the playgrounds of cyber threats. Your policy should offer guidance on best practices, like scrutinizing links, verifying sender addresses, and avoiding shady websites. It's the digital equivalent of looking both ways before crossing the street.

Keeping the Policy Clear and Accessible

Now, let's address the elephant in the room: policies are often as exciting to read as tax codes. But they don't have to be. Ditch the technical jargon and legal mumbo jumbo. Instead, write your policy in plain, straightforward language that everyone on your team can easily follow.

Finally, don't let your beautifully crafted policy collect dust in some obscure corner of a filing cabinet. Make it accessible—whether that's a shared drive, an internal wiki, or part of the onboarding process for new hires. A policy is only useful if people can find it, understand it, and apply it.

Keeping Your Cybersecurity Policy in Step with the Times

The digital world doesn't just move fast—it sprints. Threats evolve, new tools emerge, and your business adapts to the ever-changing landscape. That's why your cybersecurity policy can't be a "set it and forget it" document. It needs regular check-ups to stay relevant and effective.

Make it a habit to review your policy at least once a year. Think of it as your annual cybersecurity health check. Are there new platforms or tools your team is using? Have there been updates to regulatory requirements? Or perhaps you've encountered a new type of threat that wasn't on your radar before. Each of these shifts is a signal that your policy might need a tweak.

And when you do make changes, don't just quietly update the document and call it a day. Communicate the updates clearly to your team. Whether it's a quick announcement in a meeting, an email outlining the revisions, or a short training session to walk through the changes, ensure everyone is on the same page. After all, a policy is only as strong as the people following it.

Staying proactive with regular reviews and updates doesn't just keep your business safer—it shows your team and stakeholders that you're serious about cybersecurity. And in today's fast-paced digital world, that's a message worth sending.

Making Your Cybersecurity Policy Stick: A Team Effort

Here's the truth about policies: they're only as good as the people following them. You could write the most comprehensive, airtight cybersecurity playbook known to mankind, but if your team isn't on board, it's just words on a page. To make your policy truly effective, you need buy-in from everyone—and one of the best ways to get it is by involving your team in the process.

When employees feel like they've had a voice in shaping the policy, they're far more likely to embrace it. It's human nature—people are more committed to rules they helped create. Consider gathering input through a quick survey, holding a workshop to brainstorm best practices, or simply inviting feedback on a draft version of the policy. This isn't just a feel-good exercise; it's a practical way to identify gaps, address real-world challenges, and build a policy that makes sense for your team.

And don't stop there. Take the time to explain the "why" behind the rules. Why should passwords be long and complex? Why does clicking on unknown links pose a risk? When employees understand the reasoning, they're more likely to take the rules seriously. A cybersecurity policy isn't just about telling people what not to do—it's about empowering them to make smarter decisions.

Above all, keep your policy clear, actionable, and tailored to your business's specific needs. Nobody wants to wade through pages of technical jargon or abstract guidelines. Focus on setting clear expectations, using plain language, and providing practical steps employees can follow.

Think of your policy as the glue that binds your team together in the shared mission of protecting your business. By involving your employees and making the policy approachable, you're not just laying down rules—you're fostering a culture of security and shared responsibility. And that's the foundation of a resilient, thriving organization.

Putting It All Together: Building a Cybersecurity Game Plan

By now, it's clear that cybersecurity isn't just about firewalls and antivirus software. It's about people—your people—and the roles they play in keeping your business secure. But how do you tie all these efforts together into something cohesive? The answer lies in focusing on three foundational pillars: consistent training, a supportive culture, and a clear, actionable policy. Let's map out how these pieces work together to make cybersecurity a seamless part of your team's daily workflow.

Consistent Training: Keeping the Knowledge Fresh

Cyber threats evolve constantly, and so should your team's knowledge. Training isn't a one-and-done event—it's an ongoing process. Regularly scheduled sessions, quick refreshers, and interactive exercises keep cybersecurity top of mind and ensure everyone knows how to respond to emerging threats. Whether it's spotting phishing emails or understanding the importance of strong passwords, consistent training builds confidence and makes smart security habits second nature.

But training doesn't have to be boring. Mix in simulations, quizzes, or even a little friendly competition to keep things engaging. When learning is enjoyable, it sticks—and that means your team will be ready to act when it matters most.

A Supportive Culture: Empowering Your Team

A culture of cybersecurity starts with open communication. Employees need to feel safe admitting mistakes, asking questions, and reporting anything suspicious—without fear of blame or judgment. This is where leadership comes in. By modeling good security habits and fostering a judgment-free environment, leaders set the tone for the entire organization.

Recognition is another powerful tool. Celebrate successes, like an employee spotting a phishing attempt or following protocol in a tricky situation. These small acknowledgments reinforce positive behavior and show the team that their efforts matter.

A Clear, Actionable Policy: Setting the Ground Rules

A well-crafted cybersecurity policy is the backbone of your plan. It provides clarity, answers common questions, and establishes expectations for everyone in the organization. The best policies are straightforward, free of jargon, and tailored to the unique needs of your business.

But a policy is only effective if it's accessible and regularly updated. Make sure your team knows where to find it, and review it at least once a year to keep it current with new tools, risks, or procedures. And don't forget to involve your employees in the process—when they feel invested, they're more likely to follow the rules.

Connecting the Dots: The Power of Integration

When training, culture, and policy come together, the result is a cohesive plan that's easy to implement and maintain. Training ensures everyone knows what to do, culture creates the environment to support those actions, and the policy serves as a guiding framework. Together, these elements make cybersecurity an integral part of your team's workflow—not an afterthought.

By integrating these pillars into your business, you're not just checking off boxes; you're building a resilient organization where cybersecurity is part of the DNA. Your team will be more prepared, your data will be better protected, and you'll have the peace of mind that comes with knowing you're ready for whatever the digital world throws your way.

So, take a moment to connect the dots and map out your plan. Because when cybersecurity becomes a natural, sustainable part of your workflow, it's not just your business that benefits—it's everyone who depends on it.

Making Cybersecurity a Team Effort

Building a strong cybersecurity plan doesn't have to feel like scaling Everest. It's all about weaving smart practices into the daily fabric of your operations, making them as natural as checking emails or brewing the morning coffee. When you integrate these practices into your company's rhythm, you're not just safeguarding your business; you're transforming your team into confident, proactive defenders of your organization's digital safety.

Cybersecurity isn't a solo act—it's a team sport. Everyone has a role to play, from the intern who opens dozens of emails a day to the CEO steering the ship. When everyone works together, armed with the knowledge, tools, and culture of awareness you've cultivated, your business becomes a fortress.

With the right plan in place—clear policies, engaging training, and a culture of open communication—your team is equipped to rise to the challenge. They're not just reacting to threats; they're actively preventing them, creating a safer, more resilient environment for your business to thrive.

Remember, strong cybersecurity isn't about being perfect. It's about being prepared. And with your team standing shoulder to shoulder, ready to face the digital world's challenges, there's no threat too big to tackle together. So, keep building, keep empowering, and keep moving forward—because a team that's united in cybersecurity is a team that can handle anything.

Cybersecurity: A Mindset Built on Teamwork

Cybersecurity isn't just a set of tools or a handful of policies; it's a mindset—a collective commitment that thrives on teamwork. When everyone, from the front desk to the corner office, takes their role in protecting the business seriously, the result is more than just a secure environment. It's a united front, a powerful shield forged by vigilance and shared responsibility. And the best part? It's not about perfection or catching every single phishing attempt—it's about working together to minimize vulnerabilities and stay ahead of threats.

By empowering employees to take ownership of cybersecurity, you're doing more than just safeguarding your business. You're showing your team that their efforts matter, that their actions—whether it's reporting a suspicious email or using multi-factor authentication—are crucial pieces of a larger puzzle. Cybersecurity is no longer just the IT department's realm; it's everyone's responsibility. And when employees understand their role and feel equipped to act, they become an extension of your defenses. They catch small issues before they snowball into massive problems, becoming your business's first line of protection.

Looking Back to Move Forward

As we wrap up this chapter, take a moment to reflect on how far you've come. Building a culture of awareness and accountability isn't easy, but every effort you've made counts. Each conversation, training session, and policy update brings you one step closer to a workplace that's not just secure but resilient—a place where everyone knows the value of their contributions and takes pride in keeping the business safe.

What's Next? Building Stronger Defenses

In the next chapter, we'll layer on even more protection by diving into the technical side of cybersecurity. From robust firewalls to cutting-edge antivirus software and secure networks, we'll explore the tools and systems that give your team the support they need to succeed. These defenses, combined with the culture you've already started to build, will transform your business into a fortress—resilient, adaptive, and ready for whatever comes next.

So, let's keep moving forward, step by step, building a business that's not just prepared but proactive. Together, we're crafting a strategy that doesn't just react to threats—it anticipates them. Stronger than ever, and ready for anything.

SOCIAL ENGINEERING & HUMAN ERROR

THE HUMAN SIDE OF CYBERSECURITY

When most of us think about cyber threats, we tend to imagine shadowy figures in dimly lit rooms, hammering away at keyboards and conjuring up sophisticated code to break into systems. It's dramatic, cinematic, and—let's be honest—a little bit thrilling. But here's the real kicker: the biggest vulnerabilities in cybersecurity aren't the computers or networks at all. Nope. It's us. The humans. The people running those systems, answering those emails, and clicking those links. We're wonderfully complex creatures, sure, but we're also prone to the occasional blunder. And cybercriminals? They thrive on that.

Let's be frank: people make mistakes. We're wired to trust, to help, and sometimes, to act before we think. Maybe it's clicking on an email that looks just a tad suspicious, sharing a little too much information with a friendly stranger on the phone, or falling for a cleverly worded

"urgent" message that sends us scrambling to fix a non-existent problem. Hackers know this, and they've turned exploiting the human
factor into an art form.

Think of it like this: a business might have digital defenses that
resemble Fort Knox—firewalls standing tall, antivirus software on
patrol—but even the best technology can be undone by one unsuspecting click or one phone call where just a bit too much is shared.
Hackers don't just hack systems; they hack people. They've figured
out that it's often far easier to trick a person than to brute-force their
way through layers of digital security.

And honestly, can you blame them? Why waste hours fighting
through complex encryption when a single convincing email can deliver credentials on a silver platter? Social engineering—the hacker's
weapon of choice—uses trust, fear, urgency, and even good old-fashioned politeness to trick people into doing exactly what they want.

The statistics paint a sobering picture. Human error contributes to
the majority of cybersecurity breaches. And it's not just the rookies
making mistakes; even seasoned professionals have fallen for cleverly
disguised social engineering tactics. The point isn't to shame anyone—it's to shine a light on the very human challenges we face in the
digital world.

The reality is simple: while we can build technology as sturdy as
steel, the human element will always be the soft underbelly of cybersecurity. This chapter is all about that human element. We'll dive
into the tactics cybercriminals use to manipulate and deceive, the
psychological tricks they rely on, and, most importantly, how you and
your team can learn to spot their games before it's too late. Because
at the end of the day, the best cybersecurity system isn't just about
firewalls and antivirus software—it's about a team that knows how to
stay vigilant, think critically, and trust its instincts.

So, buckle up. Let's explore how we can outsmart the scammers, sidestep the traps, and transform human error from a vulnerability into your greatest strength.

Why People are the Weakest Link

When it comes to cybersecurity, people are often tagged as the "weakest link." Admittedly, it's not the most flattering title, but it's not because we're hopeless with technology or incapable of learning. The real issue lies in the fact that we're, well, human. We're naturally wired for connection, curiosity, and trust—all lovely traits that make us relatable, empathetic, and capable of building great relationships. Unfortunately, those same qualities can sometimes lead us a bit off course in the intricate maze of the digital world.

The Psychology of Mistakes

Let's begin with why these slip-ups happen in the first place. Picture this: you're juggling a dozen tasks, racing against a deadline, when an email pops up that seems urgent. Without a second thought—because who has time for second thoughts?—you click the link or fire off a response, just to get it out of the way. Distractions like these are a prime culprit when it comes to cybersecurity blunders. Then there are assumptions—like assuming an email is legit because it looks like it came from a familiar address. Overconfidence plays its part, too. That little voice in your head smugly declares, "I'd never fall for a scam," and before you know it, your guard is down. And let's not forget our innate desire to be helpful. Someone claims to need assistance? We're wired to want to lend a hand. Unfortunately, hackers know all too well how to exploit these very human tendencies.

How Social Engineering Takes Advantage of Trust

Social engineers—the charming con artists of the digital age—are masters of exploiting human trust. They thrive on our natural instinct to believe that people are who they say they are. Armed with a smooth demeanor and a believable backstory, they might pose as a coworker needing urgent access to a file, a vendor chasing a payment, or even an IT technician claiming to fix a problem you didn't know you had.

Their tools? Stories crafted to tug at our innate desire to be helpful or our insatiable curiosity. An email arrives with a subject line that demands your attention: "Your invoice is overdue," or "Did you request this password change?" Before you know it, you're clicking the link or responding, all because the message feels urgent—and urgency has a way of short-circuiting our better judgment. For social engineers, that moment of impulsive action is their golden ticket.

Everyday Mistakes, Big Consequences

If you've ever clicked on a suspicious link and then immediately thought, *Wait, that didn't look right...*, rest assured, you're not alone. Phishing emails remain one of the most common traps, expertly designed to catch even the most vigilant among us. And it doesn't stop there. Maybe you've shared your password with a coworker because, well, it's "just easier than setting them up with their own login." Or perhaps you've downloaded software that hasn't been officially approved by IT simply because it seemed harmless—or, let's be honest, because it was free.

These aren't acts of sabotage or rebellion; they're human moments, often driven by being busy, rushed, or simply unaware of the lurking risks. Hackers know this and count on us to make these snap decisions. The problem isn't malice—it's the pace of modern work colliding with the complexities of the digital world.

Small Mistakes, Big Consequences

Here's the thing: even the smallest mistake can swing the gates wide open for big problems. Clicking a phishing link might seem like a minor slip—until it hands a hacker the keys to your login credentials and grants access to sensitive systems. Sharing a password might feel like an innocent shortcut—until that shared password ends up in the wrong hands, and suddenly, you're playing catch-up in a crisis. The fallout from these seemingly tiny errors can be immense: unauthorized access, sensitive data leaks, financial hits, or even reputational damage so severe it takes years to mend.

It's a sobering thought, but there's a silver lining: awareness is a remarkably powerful antidote. By understanding how and why these slip-ups occur, we're better equipped to sidestep them. With the right tools, consistent training, and a dash of determination, we can flip the script. People don't have to be the weakest link—they can be the strongest line of defense.

The Art of the Con: Social Engineering's Greatest Hits

When it comes to social engineering, cybercriminals are nothing short of virtuosos at exploiting human nature. Their playbook is packed with clever tricks, each designed to capitalize on trust, curiosity, or that ever-powerful sense of urgency to nudge people into acting without thinking. The brilliance of these scams lies in their simplicity—they don't need to hack complex systems when they can hack your instincts instead.

But here's the good news: knowing the common tactics and how to spot them can give you and your team a critical edge. Let's unpack some of the top social engineering scams in the cybercriminal repertoire and, more importantly, how you can sidestep these digital landmines with confidence.

Phishing: The Cybercriminal's Favorite Bait

Phishing is the digital equivalent of those dodgy flyers stuffed into your mailbox promising miracle weight-loss pills or a free cruise. Only, instead of tempting you with questionable offers, phishing scams aim to swipe your sensitive information. These digital charades typically show up in your inbox, cleverly disguised as urgent messages from what seem to be trusted organizations—your bank, a vendor, or even your boss (and, really, who ignores an email from the boss?).

Spotting a phishing email is a bit like playing detective. The clues are there if you know where to look. Watch out for poor grammar and spelling mistakes—because, apparently, even hackers can't spell. Generic greetings like "Dear Customer" are another red flag. And if the email's tone screams urgency, with phrases like, "Your account will be locked if you don't respond immediately!"—well, that's your cue to pause and think.

The golden rule? Be skeptical. Hover over links to see where they're really taking you, and don't open attachments unless you're absolutely sure they're legitimate. A healthy dose of suspicion can be your best defense against falling for phishing bait.

Spear Phishing: Personalized Deception

While phishing casts a wide, indiscriminate net, spear phishing is a precision strike—a digital sniper shot aimed specifically at you. These cunning scams use personal details—your name, job title, or even recent activities—to make the message feel unnervingly authentic. Imagine receiving an email that says, "Hi [Your Name], I noticed you attended the recent industry conference. Here's a follow-up document from the event." Sounds harmless, right? That's precisely the trap.

The devil is in the details, and spear phishing emails are masters of detail. They're designed to lull you into a false sense of security, exploiting your trust and familiarity. That's why it's vital to take a step back and question the email before clicking or responding. Did you really meet this person? Does the email address look legitimate, or does it have telltale signs of a spoof?

When in doubt, verify. Contact the supposed sender using a method you know to be genuine, like their official website or a phone number you've used before. Don't rely on the contact information provided in the email—that's like asking a scammer if they're a scammer. Spear phishing thrives on trust, but a little skepticism can throw its aim completely off target.

Pretexting: The Art of the Con

If phishing is the quick con of the cybercrime world, pretexting is the long game. It's phishing's more sophisticated cousin, complete with a backstory and a convincing performance designed to gain your trust. Picture this: a smooth-talking "IT support" staffer calls you up, saying they're here to help reset your password, or a supposed "customer service representative" politely requests sensitive account information to resolve an issue. They've done their homework, and their story sounds plausible—almost too good to question.

But here's the catch: pretexting isn't limited to email or phone calls. It can happen in person, too, with someone showing up at your office claiming to be a repair technician or a vendor. The scam's success hinges entirely on one thing: trust. And once they've earned yours, they use it to pry open the digital doors of your business.

Avoiding pretexting requires a healthy dose of skepticism and a commitment to verification. If someone calls claiming to be from IT, politely hang up and reach out to your IT department directly using a verified number. If a visitor shows up unannounced, check with your office manager or security team before letting them access sensitive areas. As they say, "Trust, but verify." When it comes to pretexting, the extra step of confirmation is your best defense against a well-rehearsed act.

Baiting: The Freebie That Comes at a Cost

Who doesn't love free stuff? Hackers, that's who—because they know it's the perfect bait. Enter the world of baiting, where enticing offers like free software downloads, gift cards, or even a mysterious USB stick labeled "Confidential" are dangled like shiny lures. The catch? The moment you bite—whether by plugging in the USB or downloading the "free" software—you've inadvertently invited malware to set up shop on your system.

Baiting works because it taps into two universal human traits: curiosity and the thrill of getting something for nothing. That "Confidential" USB stick left conveniently on a desk? It practically begs to be plugged in. The free download promising to "double your productivity"? Who wouldn't be tempted? And hackers count on that temptation.

The antidote? A healthy dose of skepticism. If an offer seems too good to be true, it almost certainly is. Never plug in unknown devices, no matter how intriguing they look. Avoid downloading software from untrusted or unofficial sources. Remind your team that a moment of curiosity isn't worth the potential fallout of a compromised network. In the world of cybersecurity, it's better to walk away from a freebie than to pay dearly for its hidden strings.

Tailgating: The Cyber Threat on Foot

Not all cyber tricks happen behind a screen. Tailgating, or piggybacking, is a decidedly old-school tactic where an unauthorized person gains physical access to a secure area by following an employee who does have access. Picture this: someone approaches the door to your office building right behind you, balancing a coffee cup in one hand and fiddling with their phone in the other. They look busy, harmless even, and before you know it, they're slipping in on your access swipe.

It's a clever move because it preys on basic human politeness. Most people hesitate to question someone who looks the part—whether it's a courier in a hurry or someone with a confident stride who seems to belong.

But here's the thing: protecting secure areas is as important as securing networks. While it might feel awkward to question someone, it's entirely appropriate—and necessary. A simple, non-confrontational line like, "Excuse me, do you need help finding someone?" can make all the difference. If something feels off, notify security or a manager. Remember, professionalism and vigilance go hand in hand when it comes to keeping your workplace secure.

Staying one Step Ahead of Social Engineers

Social engineering scams succeed when people let their guard down, but here's the silver lining: awareness and vigilance are your best allies. Recognizing the tactics we've covered—whether it's phishing emails, pretexting phone calls, or someone tailgating their way into the office—is the first step to staying safe. The key is to pause, question, and trust your instincts.

If something feels off, it probably is. Take a moment to verify the sender of an email, double-check unexpected requests, or politely challenge unfamiliar faces in secure areas. It's not about being paranoid; it's about being cautious.

Remember, cybersecurity is more than firewalls and antivirus software. It's about empowering people—your team, your colleagues, yourself—with the knowledge and confidence to spot and stop threats before they take hold. A well-informed team isn't just a line of defense; it's the foundation of a strong, resilient organization. Stay alert, stay skeptical, and keep those instincts sharp.

Building a Culture of Cybersecurity: Small Habits, Big Impact

Creating strong cybersecurity habits in your workplace doesn't require a PhD in computer science or hours of technical training. It's more about cultivating small, intentional actions that, over time, weave a safety net across your organization. Think of it like locking the doors before leaving the office or turning off the coffee pot—simple, routine steps that make a world of difference.

These habits, while straightforward, are deceptively powerful. Each one, whether it's choosing strong passwords, double-checking email senders, or logging out of systems when stepping away, adds a layer of protection that collectively reduces vulnerabilities and minimizes risk. And the best part? They're easy to adopt and integrate into the rhythm of your workplace.

In this section, we'll explore how to embed these practices into your workplace culture—transforming cybersecurity from a daunting task into a seamless part of daily life. Because when everyone commits to these small habits, the collective impact is not just safer systems but a more confident and secure team.

Cybersecurity: Small Actions, Big Protection

When it comes to cybersecurity, consistency is your greatest ally. Think of it like locking your front door every time you leave the house—it's such a small action that you barely think about it, but its impact is enormous. The same principle applies to digital safety. Regularly practicing simple habits—like verifying the sender's email address or using unique passwords for different accounts—acts as a powerful first line of defense.

The beauty of these habits is that they don't require a degree in computer science or hours of training. They're practical, manageable steps that, when practiced consistently, can block the majority of common threats. You don't need to be a tech wizard to make a meaningful impact; you just need a routine. A little vigilance every day keeps the digital troublemakers at bay.

Pause, Think, and Outsmart Cybercriminals

In today's whirlwind of deadlines and multitasking, it's all too easy to click on a link without thinking twice or approve a request that seems urgent. Cybercriminals know this, and they count on our tendency to act first and question later. That's why encouraging a *"pause and think"* mentality in the workplace is one of the simplest and most effective ways to bolster cybersecurity. Just a moment's reflection—asking, "Does this seem right?"—can be the difference between averting an attack and dealing with a disaster.

Examples of Small Habits That Pack a Big Punch

Double-Checking Sender Information: Scammers love to exploit the tiniest details. Teach employees to scrutinize email addresses carefully—those sneaky variations, like swapping a lowercase "l" for an uppercase "I," are a favorite trick. Don't just trust the display name; click on the email address itself to confirm it's the one you expect. Cybercriminals often hide behind convincing facades, but a second glance can quickly expose their charade.

Reporting Suspicious Activity: No red flag is too small to report. Whether it's an unexpected attachment, a peculiar email, or even a slightly odd pop-up, employees should know that it's always better to speak up. Catching potential issues early is far easier than cleaning up after a breach.

Password Hygiene: Passwords are your digital keys, and strong ones are essential. Say goodbye to "Password123" and hello to unique, robust combinations. Better yet, embrace password managers to store those tricky credentials securely and eliminate the temptation to reuse passwords across accounts. Changing passwords regularly adds another layer of protection, keeping hackers perpetually locked out.

Questioning Unusual Requests: A well-crafted phishing email can make anything seem legitimate—even a request from the CEO for an urgent wire transfer. But does the CEO usually handle financial transactions via email? Probably not. Empower employees to verify unusual requests using trusted methods—like picking up the phone or walking down the hall to confirm face-to-face—before taking any action.

The Power of Small Steps

Cybersecurity doesn't have to be overwhelming. Small, thoughtful habits like these can make a huge difference in keeping your organization safe. A moment of pause here, a second look there, and suddenly the cracks that cybercriminals exploit become much harder to find. The best part? These habits, once ingrained, don't just prevent incidents—they create a workplace culture where everyone feels confident, vigilant, and part of the defense team.

Reinforcing Good CybersecurityHabits

Even the best cybersecurity habits need regular reinforcement to truly stick. Just as we're more likely to remember safety drills or workplace protocols with a little repetition, the same goes for digital safety. Here are a few creative ways to keep cybersecurity top of mind without turning it into a chore:

Posters and Visual Reminders: Sometimes a simple visual cue is all it takes. Place posters in high-traffic areas like break rooms or near computer stations with quick, punchy reminders such as "Think Before You Click" or "Verify Before You Share." These serve as gentle nudges to keep employees alert in their day-to-day interactions.

Monthly Cybersecurity Tips: A short, actionable tip delivered via email or chat can work wonders for maintaining awareness. Think along the lines of, "Check sender email addresses carefully—scammers often rely on sneaky lookalike addresses to fool you," or "Update your passwords this week to keep accounts secure!" Quick, digestible nuggets of advice keep the topic fresh without overwhelming your team.

Team Meeting Check-Ins: Carve out five minutes during regular team meetings to talk about recent security tips or share "near-miss" stories (anonymously, if needed). For instance, recounting how someone almost clicked on a phishing email but caught it just in time reinforces the importance of vigilance. These discussions not only make cybersecurity relatable but also foster a culture of open communication and shared responsibility.

Make It a Team Effort

By weaving cybersecurity reinforcement into the daily rhythm of work, you ensure it becomes part of your organization's culture, not just a box to tick. These small, consistent reminders empower your team to stay sharp and proactive, making the entire business stronger in the face of digital threats.

The Power of Small Habits

By honing in on small, manageable habits, you're creating a workplace where cybersecurity feels less like a daunting task and more like an everyday routine. These simple practices—double-checking email addresses, pausing before clicking links, and using strong passwords—don't just protect your systems. They empower your people.

Over time, these habits become second nature, like buckling your seatbelt or locking the door behind you.

What's more, cybersecurity isn't just about firewalls, antivirus software, or the latest tech buzzwords. It's about people. It's about your team taking consistent, thoughtful actions to safeguard themselves, their colleagues, and the organization as a whole. These small, steady efforts add up to a culture of vigilance that strengthens your defenses from the inside out.

Cybersecurity may feel like a complex puzzle, but at its heart, it's about people making good choices, one small habit at a time. And those choices? They're what make all the difference.

Case Studies: Real-World Examples of Social Engineering Attacks

Social engineering attacks aren't about outsmarting machines—they're about outsmarting people. These crafty cybercriminals bypass firewalls and antivirus software by targeting the one vulnerability that technology can't patch: human nature. Trust, curiosity, and the innate desire to help are wonderful qualities, but in the hands of a skilled scammer, they can quickly become tools of exploitation.

Let's take a look at three real-world scenarios that show just how deceptively simple these attacks can be—and the vital lessons we can learn to stay one step ahead.

Case Study 1: The Fake CEO Scam

The Scenario: It's 4:45 p.m. on a Friday, and Sarah, an accounts payable specialist, is wrapping up for the week when an email lands in her inbox. The subject line grabs her attention: "URGENT: Need Your Help." The sender? The company's CEO.

In the email, the CEO explains that they're stuck in an important meeting and need Sarah to wire $50,000 to a supplier immediately. There's a sense of urgency—borderline panic—in the tone, with phrases like, "Please handle this right away to avoid delays." Wanting to be helpful and avoid letting the CEO down, Sarah acts quickly and processes the payment.

Come Monday morning, the real CEO stops by to discuss an unrelated matter. That's when the truth hits: the email was a scam. The CEO never requested the payment, and the money was wired to a fraudulent account overseas.

What Went Wrong:

- Sarah didn't stop to verify whether the email was authentic.

- The urgency of the request pressured her to act quickly, bypassing the company's standard procedures.

- The email contained subtle red flags that were easy to miss under pressure—like a slightly misspelled email address and generic phrasing.

What Could Have Been Done Differently:

- **Verify Unusual Requests:** A quick phone call to the CEO or a trusted manager could have immediately confirmed the email was fake.

- **Implement a Two-Step Process:** For large financial transactions, having a second approver or requiring verbal confirmation adds a crucial layer of protection.

- **Pause and Think:** Encourage employees to take a moment to double-check urgent or unusual requests, especially those involving money.

The Lesson: Cybercriminals know that urgency clouds judgment. By fostering a culture where employees feel empowered to slow down and verify requests, businesses can avoid falling victim to these cleverly disguised cons. A quick pause on Friday could have saved Sarah—and the company—a major headache come Monday.

Case Study 2: The "IT Support" Pretexting Attack

The Scenario: Mark, a marketing manager, is deep in the weeds on a big project when he gets a call from someone claiming to be from the company's IT department. The caller sounds polished and professional, referencing an internal IT ticket number and even dropping the name of someone on the actual IT team. They explain there's an issue with Mark's email account and, unfortunately, they need his username and password to fix it.

Mark hesitates, but the caller's tone is reassuring, and the details seem legitimate. Not wanting to delay the fix—or risk losing access to his account—Mark shares his credentials over the phone.

Fast-forward a few days, and Mark notices odd activity in his email account. Colleagues are replying to messages he didn't send, and some are confused by phishing emails supposedly coming from him. By the time IT steps in, the damage is done. The hacker used Mark's credentials to access sensitive company files and launch phishing attacks targeting clients.

What Went Wrong:

- Mark didn't verify the caller's identity before sharing his credentials.

- He didn't question why IT would need his password to fix an internal issue—a major red flag.

What Could Have Been Done Differently:

- Never Share Passwords: Employees should know that passwords are private and should never be shared over the phone, email, or any other channel.

- Verify Callers: If someone claims to be from IT, hang up and call back using an official company number to confirm their identity.

- Clear IT Policies: IT departments should emphasize policies like *"We will never ask for your password"* to help employees recognize potential scams.

The Lesson: Hackers rely on confidence and clever storytelling to exploit trust. By training employees to verify requests for sensitive information and fostering a culture where it's okay to pause and double-check, companies can significantly reduce their risk. A quick call to IT for verification could have stopped this scam in its tracks.

Case Study 3: Baiting with a USB Drive

The Scenario: On her way into work one morning, Lisa spots a USB drive lying in the parking lot, labeled enticingly as "Confidential Payroll Data." Her curiosity piqued, she picks it up and decides to check what's on it.

Back at her desk, she plugs the drive into her computer. Almost immediately, strange things start happening: pop-ups flood her screen, files disappear, and her computer slows to a crawl. Unbeknownst to Lisa, the USB drive wasn't full of payroll data—it was loaded with malware designed to give hackers remote access to her system.

What Went Wrong:

- Lisa's curiosity led her to overlook the potential risks of plugging in an unknown device.

- She wasn't aware of (or didn't follow) any company policies about handling suspicious devices.

What Could Have Been Done Differently:

- Avoid Unknown Devices: Employees should be trained never to insert unverified USB drives into their computers, no matter how tempting the label.

- Establish a Policy: A clear procedure should exist for dealing with found devices, such as turning them over to IT for secure examination.

- Raise Awareness: Employees should understand that seemingly innocuous items, like USB drives, can be weaponized by hackers through a tactic known as "baiting."

The Lesson: Hackers often rely on curiosity as a weak point in human behavior, and baiting attacks like this are designed to exploit it. By educating employees about these risks and establishing clear guidelines, businesses can prevent what may seem like a harmless action from spiraling into a full-scale cyberattack. A simple "If you don't know where it came from, don't plug it in" rule could have saved Lisa—and the company—a lot of trouble.

Lessons Learned from Each Case

1. **Awareness is Key:** Each of these examples demonstrates how social engineering exploits human tendencies like trust, curiosity, and urgency. The more employees understand these tactics, the better equipped they are to recognize and resist them. Awareness is the first line of defense.

2. **Verify, Don't Assume:** Whether it's an urgent email from the "CEO" or a call from "IT," verification is essential. Taking a moment to double-check a request—by calling the

supposed sender or consulting a colleague—can save a business from costly errors. Trust is good, but trust with verification is better.

3. **Stick to Policies:** Clear, well-communicated policies are the backbone of a secure organization. From how to handle financial transactions to the dos and don'ts of password sharing or dealing with unknown devices, guidelines ensure everyone knows the correct actions to take. Regular training ensures those policies are followed, not forgotten.

4. **Encourage a Culture of Caution:** No one should feel embarrassed to ask questions or report suspicious activity. A culture that values caution over compliance reduces hesitation, empowers employees, and fosters vigilance. The goal is a team that feels supported in making the right call—even if it means asking twice or saying no to unusual requests.

The Big Picture

These lessons aren't just about addressing past mistakes—they're about preventing future ones. By fostering skepticism and careful decision-making, businesses can create a workplace that doesn't just react to social engineering threats but actively resists them. Social engineers rely on people letting their guard down, but with the right training and culture, employees can become the strongest link in the cybersecurity chain.

Building a People-Centered Security Plan

Creating a security plan that places people at its core means understanding that cybersecurity isn't just a technical problem—it's a human one. While firewalls and antivirus software are vital, they're only as effective as the people using them. Cybersecurity lives in the day-to-day actions of your team: how they handle emails, how they stay informed, and, most importantly, how they react when something feels off.

Here's how to build a security plan that turns your greatest asset—your team—into your most robust line of defense:

1. **Start with Awareness:** Educate your employees about common threats like phishing, social engineering, and malware. Awareness training helps them recognize red flags and equips them with the knowledge to navigate potential threats confidently.

2. **Emphasize Accountability:** Make it clear that everyone has a role to play in protecting the organization. From the receptionist to the CEO, cybersecurity is a shared responsibility. Regularly remind your team that their actions matter and contribute to the company's overall safety.

3. **Encourage Vigilance:** A strong security plan empowers employees to speak up. Create a culture where it's second nature to report anything suspicious, whether it's a strange email, an unrecognized phone call, or an unexpected pop-up. Quick reporting can stop a threat in its tracks.

4. **Reinforce Good Habits:** Small, consistent actions—like verifying email senders, using strong passwords, and following secure data handling practices—add up. Build routines that make these habits feel as natural as locking a door or fastening a seatbelt.

5. **Support with Technology:** While the focus is on people, the right tools make their job easier. Equip your team with resources like password managers, multi-factor authentication, and secure communication platforms to reduce friction and strengthen security practices.

By making your team the centerpiece of your cybersecurity strategy, you're not just mitigating risks—you're empowering your employees to take ownership of the company's safety. Cybersecurity isn't just about keeping the bad guys out; it's about building a workplace where everyone feels confident, informed, and ready to respond. When your people are engaged and prepared, they become the most vital part of your defense plan.

Cybersecurity as a Daily Habit

Cybersecurity isn't something that should live on the fringes of your business—it needs to be woven into the very fabric of daily operations. Think of it like workplace safety: just as everyone instinctively avoids leaving tripping hazards in the hallways, everyone should also develop the reflex to practice basic cybersecurity habits. It's the kind of thing that, once ingrained, becomes second nature but carries enormous benefits for the entire organization.

Let's revisit some of these habits, because they bear repeating. Locking your computer when you step away from your desk? A no-brainer. Pausing to scrutinize a suspicious link before clicking? Absolutely essential. These small, everyday actions may seem trivial on their own, but collectively, they create a ripple effect of protection that can shield your business from significant threats.

And here's the kicker: fostering this mindset starts with leadership. When leaders make cybersecurity a priority and model good practices, it sends a powerful message to the entire team. Whether it's casually mentioning security tips during a team meeting, sharing an interesting article about the latest phishing scams, or simply being vocal about the importance of vigilance, leaders set the tone.

When cybersecurity becomes part of the regular conversation—like discussing project updates or celebrating wins—it shifts from being "the IT department's job" to a shared responsibility. Over time, this collective effort transforms your organization into a safer, more resilient place, where everyone understands that their role, however small it may seem, contributes to the overall security of the business.

Cybersecurity is Always Changing

If there's one constant in cybersecurity, it's that nothing stays the same for long. Hackers adapt, threats evolve, and yesterday's best practices can quickly become outdated. That's why cybersecurity training can't be a one-and-done affair—it needs to be revisited regularly to keep pace with the ever-shifting landscape of digital threats.

Think of it as brushing up on your skills, like renewing a first aid certification or updating workplace safety protocols. Scheduling short, focused training sessions every quarter—or whenever new threats emerge—keeps your team sharp and aware.

And don't worry—these sessions don't need to be daunting marathons. The best training is practical and to the point. Cover real-world scenarios like spotting phishing attempts, crafting strong passwords, or recognizing the telltale signs of social engineering. When training is relevant, clear, and easy to understand, it gives employees the confidence to apply what they've learned immediately.

The goal here isn't to transform your team into cybersecurity wizards capable of writing code or battling malware in real time. It's simply to equip them with enough knowledge and awareness to recognize when something's amiss and know what steps to take. When your team feels prepared, they're less likely to panic—and far more likely to act effectively—in the face of potential threats. It's about creating a workplace culture where cybersecurity feels approachable, actionable, and, above all, doable.

Mistakes Happen, but Silence Is Costly

Let's face it—nobody's perfect. Mistakes are as human as coffee spills and typos. But in the realm of cybersecurity, the real danger often isn't the mistake itself—it's the hesitation to admit it. Picture this: an employee clicks on a phishing email, panics, and then stays silent, worried about repercussions. That silence can be a hacker's golden hour, offering them time to exploit the breach while the organization remains blissfully unaware.

The key to avoiding such scenarios? Foster a culture where honesty trumps fear. Employees should feel comfortable reporting issues—whether it's a suspicious email, a strange pop-up, or even their own accidental slip. Leadership sets the tone here. A response of gratitude and support, rather than reprimand, not only encourages transparency but also reinforces the idea that mistakes are opportunities to learn and grow.

Think of it this way: a judgment-free culture isn't about ignoring accountability—it's about creating an environment where accountability can thrive. When employees trust that they won't be punished for speaking up, they're far more likely to report problems promptly. And in cybersecurity, those precious moments of early reporting can mean the difference between a contained incident and a full-blown crisis.

Encourage your team to treat mistakes as stepping stones rather than stumbling blocks. After all, the strongest defenses are built on a foundation of trust, transparency, and a willingness to learn from missteps.

Making Cybersecurity Fun: Gamification to the Rescue

Who says cybersecurity training has to be all spreadsheets and sighs? With a dash of creativity and a touch of competition, you can turn what might otherwise feel like a technical slog into something genuinely enjoyable. Enter gamification—a way to make learning about cybersecurity not just tolerable, but dare I say, fun.

Imagine kicking off a company-wide trivia quiz focused on phishing scams, with a small prize (or just bragging rights) for the highest scorer. Or perhaps you create a "Spot the Fake" challenge, where employees try to sniff out suspicious emails or messages, racing against the clock. For an added layer of camaraderie, why not hold a team competition? Departments could compete to see who finishes the most training modules or scores the highest on quizzes.

Gamification doesn't just lighten the mood—it taps into our natural competitive streak and transforms learning into a shared experience. It helps reinforce good habits in a way that sticks because people are actively engaged rather than passively enduring. Plus, who doesn't want the chance to one-up their coworkers in the name of cybersecurity?

A Human-Centric Approach: Empowerment Through Education

When you treat cybersecurity as a team sport—where everyone has a role, and learning is baked into the workplace culture—you're not just safeguarding your business. You're empowering your team to take ownership of security in a way that feels both meaningful and manageable. A human-centric security plan doesn't just reduce risks; it creates an environment where employees feel valued, proactive, and confident in their contributions.

By making learning enjoyable, weaving education into daily operations, and focusing on collaboration, you're building more than just a defense against cyber threats—you're cultivating a culture of resilience and teamwork. And that, as they say, is a win-win for everyone.

Cybersecurity: Turning Vulnerabilities into Strengths

When it comes to cybersecurity, people are a paradox. They're both your greatest vulnerability and your greatest strength. But here's the uplifting part: awareness changes everything. When employees understand the tactics behind social engineering and know what warning signs to look for, they transform from potential liabilities into your most reliable first line of defense.

Awareness isn't just about knowing that threats exist—it's about building the confidence to recognize them and the resolve to stop them in their tracks. Imagine your team not just reacting to suspicious activity but proactively guarding against it. That's the power of an informed and vigilant workforce.

But, of course, awareness alone won't cut it. Like any skill worth having, good cyber habits need consistent reinforcement. It's the small, seemingly mundane actions—pausing before clicking a link, scrutinizing the sender's email address, or verifying a request—that make all the difference. These aren't grand gestures, but when practiced daily, they become second nature, seamlessly woven into the fabric of your business. Over time, these small habits evolve into a culture of cybersecurity that becomes as automatic as flipping off the lights when you leave a room.

Building Resilience, One Step at a Time

As we close this chapter, take a moment to appreciate the progress you're making. By tackling the human side of cybersecurity, you're laying the foundation for a stronger, more resilient business. But people, while critical, are just one piece of the puzzle.

In the next chapter, we'll shift our focus to the tools and technologies that can work in harmony with your team's efforts. From firewalls to multi-factor authentication, these digital allies can fortify your defenses and ensure that your organization is prepared for whatever comes its way. Together, we'll build a multi-layered strategy that protects what matters most—your data, your team, and your business.

Stay tuned—there's more to come, and it's going to be good.

THE ROLE OF REGULATIONS &
COMPLIANCE

KEEPING IT LEGAL & SECURE

Introduction to Cybersecurity Regulations and Compliance

When you hear the phrase "cybersecurity regulations," your mind might jump straight to visions of endless paperwork, stacks of forms, and bureaucratic red tape. But let's hit pause on that thought for a moment. The truth is, these regulations aren't just about ticking boxes—they exist to protect you, your business, and the people who trust you with their data. That's right: your customers, employees, and partners are all counting on you to get this right.

Think of compliance as more than a chore to avoid fines. It's about building trust and credibility in an increasingly security-conscious world. And let's not gloss over those fines and consequences, either. Failing to comply with cybersecurity regulations can cost you—big time. We're talking about hefty penalties, potential legal headaches, and a PR nightmare if a data breach puts your customers' personal information out in the wild. Imagine trying to explain to your loyal clientele that their data was leaked because your business didn't meet basic security standards. It's the kind of conversation no one wants to have—and compliance is your best defense against that scenario.

The importance of data privacy and security has never been more evident. High-profile breaches dominate the headlines, and consumers are paying closer attention than ever to how their information is handled. Whether you're running a cozy local coffee shop or managing a growing online empire, the spotlight is on. The days of thinking, "We're too small to be a target," are long gone. Accountability isn't just for the big players anymore—it's for everyone.

Of course, if you've ever tried to wade through the jargon of cybersecurity regulations, you'll know it can feel like stepping into a foreign language class you didn't sign up for. GDPR, HIPAA, PCI DSS—it's an overwhelming alphabet soup that can make even the most organized business owner's head spin. But here's the good news: it's manageable. You don't have to tackle it all at once, and you don't need a legal dictionary at your side.

In this chapter, we'll strip away the complexity and break down what you really need to know. We'll explore why these regulations matter, how they affect your business, and how you can stay compliant without pulling your hair out. Think of this as your roadmap to navigating the legal landscape of cybersecurity. Together, we'll make sense of it all and ensure your business is not just compliant, but confident.

Let's dive in.

The Landscape of Cybersecurity Regulations: Which Ones Matter to You?

When it comes to cybersecurity regulations, it's not a question of whether they apply to your business—it's a matter of figuring out which ones do. These aren't just arbitrary hoops to jump through; they're designed to protect sensitive data and safeguard trust between your business and your customers. Let's break it down into digestible pieces so you can make sense of what these regulations mean for you.

General Data Protection Regulation (GDPR)

Let's kick things off with GDPR. Originating in the European Union, this regulation has a global reach. If your business collects or processes data from EU citizens—even if you're on the opposite side of the planet—GDPR applies to you.

At its core, GDPR is about giving individuals control over their personal data. The key requirements? You'll need crystal-clear policies on how you're using and protecting data, explicit consent from users before collecting it, and a plan to notify authorities (and affected individuals) if a breach occurs.

Think of GDPR as the gold standard for data privacy—strict but designed to foster trust and accountability. It's not just about compliance; it's about showing customers that their privacy is a priority.

Health Insurance Portability and Accountability Act (HIPAA)

If you're in the healthcare industry—or even tangentially involved, like handling medical billing or IT support for a clinic—HIPAA is your bible. This regulation ensures that protected health information (PHI) is stored securely and shared only with proper authorization.

Compliance involves encrypting medical data, training employees on privacy protocols, and securing patient consent before sharing information. It's all about maintaining confidentiality for some of the most sensitive data out there—health records.

Payment Card Industry Data Security Standard (PCI DSS)

If your business handles credit card payments, PCI DSS is a must-know. Whether you're running a corner café or a bustling on-line shop, this regulation is your guide to keeping payment card data secure.

The requirements? Protect your network, control who can access sensitive payment information, and conduct regular vulnerability checks. Credit card data is a prime target for hackers, and PCI DSS is your shield against becoming an easy mark.

Other Industry-Specific Regulations

Certain industries or locations come with their own additional rule-books.

- **Financial Services**: If you work in this sector, FINRA provides guidelines for protecting sensitive financial information.

- **California Consumer Privacy Act (CCPA)**: If you handle data from California residents, you'll need to meet strict standards for transparency and security.

Each regulation comes with its own quirks, but they all share a common goal: protecting sensitive information and maintaining trust.

The Common Threads Across Regulations

While each regulation has its own unique requirements, they all boil down to a few universal principles:

- **Data Protection**: Whether it's health records, credit card numbers, or personal data, keeping it secure is non-negotiable.

- **Breach Reporting**: If something goes wrong, you're expected to own up to it quickly and notify the appropriate parties. Transparency is critical.

- **User Consent**: From GDPR to CCPA, regulations increasingly emphasize the right of individuals to control how their data is used.

Compliance: Not Just a Box to Check

While these regulations might seem overwhelming at first, they're about more than just avoiding fines or legal trouble. Compliance is about building trust—a solid foundation that reassures your customers, employees, and partners that their data is in safe hands.

It's not just good legal practice; it's good business. By meeting these standards, you're not only safeguarding your operations but also strengthening relationships with the people who matter most to your success. So, let's dive into the specifics and take this one step at a time.

Governments Step into the Cybersecurity Arena

Cybersecurity is no longer just a challenge for businesses trying to outwit hackers in their spare time. As digital threats grow more sophisticated, governments worldwide are stepping up their game, creating regulations, guidelines, and enforcement mechanisms to tackle vulnerabilities head-on and protect their citizens.

But what does this increased oversight mean for your business?

First, it's a signal that cybersecurity isn't optional—it's a public concern that affects everyone, from individuals scrolling on their phones to multinational corporations managing terabytes of sensitive data. Second, it means your business will likely face new expectations, not just from regulators but from customers and partners who want to know their data is in safe hands.

So, whether you're a small retailer or a mid-sized manufacturing firm, these regulations aren't just noise in the background. They're a roadmap to better digital hygiene and a chance to prove that you take security seriously.

In this section, we'll break down how government initiatives and regulations are shaping the cybersecurity landscape and what they mean for businesses of all shapes and sizes. Spoiler alert: understanding these changes isn't just about staying compliant; it's about staying competitive.

Increased Government Oversight in Cybersecurity

It's no secret—governments are watching. With the rise in high-profile cyberattacks and data breaches, regulators are stepping up their game. They're enacting stricter laws, demanding transparency, and holding businesses accountable when things go wrong. This might sound intimidating, but the goal is to create a safer digital environment for everyone.

Stricter regulations often include requirements for businesses to implement robust cybersecurity measures, disclose breaches promptly, and ensure sensitive customer data is handled securely. Fines for non-compliance can be hefty, and enforcement is no longer just aimed at major corporations—small and medium-sized businesses are under the microscope too.

Government Take the Lead in Cybersecurity

We've seen some impressive moves by governments in recent years to tackle the growing menace of cyber threats. These actions aren't just bureaucratic box-ticking; they're shaping how businesses worldwide handle security and data privacy. Let's take a look at a few examples:

- **General Data Protection Regulation (GDPR)**: The European Union's GDPR didn't just set the bar for data privacy—it raised it to skyscraper levels. This regulation demands that companies be crystal clear about how they collect, use, and store personal data. It also comes with teeth in the form of hefty fines for breaches, reminding businesses everywhere that privacy isn't just a courtesy; it's a legal obligation.

- **California Consumer Privacy Act (CCPA)**: Across the pond, California decided to get in on the action with CCPA, giving U.S. consumers more control over their data. Businesses must now handle personal information with care, ensuring it's secure and accessible for review or deletion at the customer's request. Accountability is the name of the game here.

- **Cyber Incident Reporting Requirements**: Many countries have introduced mandatory reporting rules for cyberattacks. Companies are now required to alert national agencies within a tight timeframe when an incident occurs. These quick reports allow authorities to respond faster and limit the damage from ongoing attacks.

These aren't abstract ideas floating around in government offices—they come with real-world consequences. High-profile cases, like multi-million-dollar fines slapped on tech giants for mishandling user data or breaches that spilled millions of personal records, show just how serious regulators are about enforcement. The message is clear: cybersecurity isn't optional. It's a must-have for businesses that want to survive and thrive in a digital-first world.

Watchdogs and Allies: The Dual Role of Regulatory Agencies

Regulatory bodies and cybersecurity agencies are stepping up in a big way, serving as both watchdogs and allies in the ever-evolving digital landscape. On one hand, they're the enforcers, ensuring businesses play by the rules; on the other, they're also a guiding hand, offering resources to help businesses stay ahead of threats.

Take the **Federal Trade Commission (FTC)** in the U.S. or the **Information Commissioner's Office (ICO)** in the U.K., for example. These agencies have become the guardians of consumer protection in the digital age. They investigate data breaches, enforce compliance with cybersecurity regulations, and, when necessary, impose hefty penalties on those who fall short. If you've ever heard of a company paying millions after a data mishap, chances are one of these agencies was behind it.

But it's not all about penalties and investigations. These agencies also want businesses to succeed in staying secure. Programs like the **Cybersecurity and Infrastructure Security Agency (CISA)** in the U.S. and the **European Union Agency for Cybersecurity (ENISA)** offer a treasure trove of resources. From detailed guidelines to practical tools and even training programs, they equip businesses with the knowledge and support needed to fend off cyber threats.

In essence, these organizations play both roles with finesse: the stern disciplinarian keeping everyone in check and the helpful mentor ensuring businesses have what they need to thrive in a digital world.

No Businesses is Too Small for Cybersecurity

If you're thinking, *"I'm just a small business—this doesn't apply to me,"* it's time to rethink that stance. Hackers don't just target the big fish; in fact, they often aim for small businesses precisely because they assume your defenses are weaker. And governments and regulators? They know this, which is why compliance applies across the board, from multinational corporations to the local coffee shop.

Here's the deal: if you collect customer data—whether it's credit card numbers, email addresses, or even just a name and phone number—you're on the hook for protecting it. Cybersecurity regulations aren't just about keeping governments happy; they're about safeguarding the trust that consumers place in you. Failing to comply doesn't just risk fines and penalties. It risks your reputation, your customers' loyalty, and, potentially, your entire business.

But here's the silver lining: staying ahead of regulations doesn't have to feel like scaling a mountain. Stick to the cybersecurity basics, like using strong passwords, encrypting sensitive information, and regularly updating your software. Set aside time to review your practices and ensure they align with current standards.

Being proactive about compliance isn't just about dodging fines. It's about showing your customers that you value their trust. Every step you take to secure their data is a step toward building a stronger relationship with them—and a more resilient business for yourself.

Staying Ahead in an Ever – Changing cybersecurity Landscape

The landscape of cybersecurity is like a constantly shifting sand dune. Just when you think you've got it figured out, the winds change, and everything looks different. If there's one thing we can count on, it's that the rules of the game will keep evolving. Cyber threats are becoming more sophisticated, global regulations are tightening, and the need for stronger defenses is only growing.

For small businesses, staying ready doesn't mean trying to predict every twist and turn. The goal is to prepare for the unknown. By staying informed about emerging threats and adapting to new regulations, you can position your business to meet the challenges ahead.

Let's take a peek into what's on the horizon and how you can navigate these changes without losing your footing. Because in the ever-shifting world of cybersecurity, being ready is half the battle.

The Push for Global Cybersecurity Standards

Cyber threats are the ultimate equalizers—they don't care about geography, time zones, or the size of your business. This indifference has sparked a growing realization among governments and international organizations: a fragmented approach to cybersecurity simply won't cut it. To truly tackle the rising tide of digital threats, global cooperation is becoming more of a necessity than a lofty ideal.

The push for unified global cybersecurity standards is gaining momentum, and while it might sound like an ambitious undertaking, progress is already being made. These initiatives aim to create cohesive frameworks that transcend borders, making it easier for businesses to navigate the labyrinth of regulations.

For small businesses, this shift holds promising potential. Instead of grappling with a patchwork of rules that vary from one country to the next, global standards could provide a more consistent and straightforward playbook. Whether your customers are local or scattered across continents, clear and unified guidelines mean less confusion and more focus on building strong defenses.

It's a hopeful step toward simplifying compliance while strengthening collective security—a win-win for businesses of all sizes.

The Growing Demand for Data Privacy

If you've noticed that data privacy is dominating the conversation lately, you're not alone. Consumer awareness around how personal information is handled has skyrocketed, ushering in major regulations like GDPR in the EU and CCPA in California. But here's the kicker—this is just the beginning.

As consumers continue to demand more control over their personal data, the regulatory landscape is expanding. Countries, regions, and even local governments are jumping on board with their own data privacy laws. For businesses, this means the era of following just one or two big-name regulations is over.

For small businesses, this growing patchwork of rules presents both a challenge and an opportunity. On the one hand, keeping up with evolving laws that might affect your industry, customer base, or geographic footprint requires diligence. On the other hand, demonstrating a commitment to data privacy isn't just about ticking legal boxes—it's about building trust with your customers.

The takeaway? Data privacy isn't a passing trend or a niche concern. It's becoming the new normal, and staying ahead of the curve is as much about safeguarding your reputation as it is about staying compliant.

Preparing for Tomorrow's Regulations

As technology marches on, so do the threats that come with it—and the regulations aimed at keeping them in check. Looking ahead, we can expect a wave of new rules tackling some of the most cutting-edge areas of innovation.

Artificial Intelligence (AI)

AI is quickly becoming a cornerstone of business operations, from chatbots to predictive analytics. But with great power comes great responsibility—and likely, new regulations. Future rules will likely focus on ensuring the ethical use of AI, securing the vast amounts of data these systems rely on, and mitigating the risks of AI-driven decisions going awry.

Internet of Things (IoT)

The Internet of Things has transformed everything from how we manage office equipment to how we secure our spaces. But with more devices connected, there are more opportunities for vulnerabilities. Regulators are expected to zero in on IoT security, ensuring that smart devices don't inadvertently open the door to cyber threats.

Biometric Data

Fingerprint scans, facial recognition, voice authentication—biometric data is becoming more common in both personal and professional spaces. But with its sensitivity and permanence, storing and protecting this data will likely come under tighter scrutiny. Future regulations will focus on safeguarding biometric information to ensure it's not misused or exposed.

Staying Ahead of the Curve

Complying with these future regulations won't just mean reacting to changes as they come—it'll require proactive thinking. Keeping your business secure will involve staying informed about emerging technologies, anticipating potential vulnerabilities, and adapting your security practices accordingly.

The bottom line? As technology evolves, so must your approach to compliance and security. By keeping one eye on the horizon and the other on your day-to-day operations, you'll be ready to adapt, innovate, and stay ahead in a rapidly changing digital landscape.

Staying Ahead Without Feeling Overwhelmed

Let's be honest—keeping up with evolving regulations can feel a bit like chasing a moving target. But here's the good news: preparing for what's next doesn't have to be a Herculean task. With some small, consistent efforts, you can stay ahead of the curve and save yourself from scrambling when new rules come into play.

1. Review Your Data Practices Regularly

Think of this as a cybersecurity tune-up. Regularly assess how your business collects, stores, and uses data. Is everything encrypted? Are access permissions still appropriate? Identifying potential weak spots now will save you a headache (and possibly a hefty fine) down the road when regulations inevitably demand stricter measures.

2. Keep an Eye on Trends

Staying informed doesn't mean you have to dive into endless legal jargon. Follow newsletters, blogs, or social media feeds from reputable cybersecurity organizations, regulators, or industry peers. These sources distill the essentials, keeping you in the loop about upcoming regulations and industry trends without bogging you down in complexity.

3. Invest in Adaptability

The best defense against an unpredictable regulatory future is a flexible foundation. This means training your team to stay security-conscious, implementing scalable cybersecurity tools, and ensuring your policies are living documents that can evolve with the times. Being adaptable now makes pivoting later far less daunting.

Looking to the Horizon

It's easy to see new regulations as just another hurdle to clear, but remember: the ultimate goal isn't to make your life harder. These laws aim to safeguard businesses like yours and protect the customers who trust you with their data. By staying proactive and informed, you're not just ticking compliance boxes—you're positioning your business to thrive in an increasingly complex digital world.

Think of it as future-proofing your operations. Each small step you take today—whether it's tightening data practices or training your team—builds a safer, more resilient future for your business.

Let's Fact it—Compliance Can Be Overwhelming

Compliance often feels like diving headfirst into an endless labyrinth of rules and regulations, complete with acronyms that sound like they belong in a spy thriller. GDPR, HIPAA, PCI DSS—it's like someone handed you a decoder ring and then walked away without explaining how it works.

But here's the silver lining: staying compliant doesn't have to be a Herculean task that derails your day-to-day operations. With a clear strategy and a bit of proactive effort, you can navigate this maze without losing your sanity—or your productivity.

Breaking It Down

The key is to break compliance into bite-sized, manageable steps. Think of it like assembling a puzzle: tackle one piece at a time, and suddenly, the bigger picture starts to make sense. Instead of trying to overhaul your entire system at once, focus on building a plan tailored to your business's needs.

This approach not only saves time but also reduces the stress that comes with trying to juggle regulations while running your business. With the right structure in place, compliance can become less of a headache and more of an integrated, almost invisible part of your operations.

So, let's demystify the process together, one step at a time, and transform compliance from a daunting chore into a sustainable part of your business strategy. Because protecting your data and meeting regulations shouldn't feel like a punishment—it's simply good business sense.

Think of Compliance as Building a House

Your compliance framework is like the foundation of a house—solid, reliable, and absolutely critical for keeping everything upright. But you don't need to construct a mansion overnight. Start small, lay a few bricks at a time, and before you know it, you'll have a structure that stands strong against whatever comes its way.

Step 1: Identify Your Requirements

Begin by figuring out which regulations apply to your business. Are you handling sensitive healthcare data (HIPAA)? Processing credit card payments (PCI DSS)? Collecting information from EU citizens (GDPR)? Knowing your specific obligations is like drawing up blueprints—it's the first step toward building something sturdy.

Step 2: Create a Compliance Checklist

Think of this as your construction plan. Write down the key actions your business needs to take. Whether it's setting password policies, securing data storage, or putting a breach notification process in place, having a clear list ensures you don't miss any crucial steps.

Step 3: Assign Responsibilities

No house is built by one person, and compliance shouldn't rest on a single set of shoulders either. Assign tasks to team members. Perhaps one person oversees data protection while another handles incident reporting. Dividing responsibilities not only lightens the load but ensures accountability.

Step 4: Review Regularly

Houses need maintenance, and so does your compliance framework. Regulations change, and your business evolves. Schedule regular check-ups to make sure everything is still up to code.

Keep It Simple

You don't need to write a 100-page compliance manual overnight. Focus on small, intentional steps—like making sure everyone understands their role or ensuring your data storage is secure. The goal is progress, not perfection, and every little step brings you closer to a resilient, compliant foundation for your business.

Know Your Risks to Reduce Them

Compliance boils down to one simple idea: reducing risks. But before you can tackle them, you have to figure out where they're hiding. Think of a risk assessment as a treasure hunt—not for gold, but for the weak spots in your defenses. And the good news? It doesn't have to be overly complicated. Here's how to break it down:

1. Assess Your Data

Start by taking stock of the sensitive information your business handles. What kind of data do you collect—financial details, customer records, intellectual property? Where is it stored, and who has access to it? Knowing the answers to these questions is like mapping out the treasure (or in this case, the targets hackers might go after).

2. Check Access Controls

Now that you know where your data lives, make sure it's not wide open for anyone to stroll in. Confirm that only authorized personnel can access sensitive information. This is especially crucial for data like credit card numbers, employee records, or proprietary business strategies.

3. Review Data Storage Practices

Take a close look at how and where your data is stored. Is it encrypted to keep prying eyes out? Are backups tucked away safely, preferably offsite or in the cloud? Good storage practices are like locking up the family jewels—they don't just lie around waiting for trouble.

4. Test Incident Response Plans

Here's where it gets interesting: stage a mock breach. Pretend a hacker got in—what would your business do next? Role-playing these scenarios is like running a fire drill for your digital house, exposing gaps in your defenses before the real thing happens.

Prepared, Not Perfect

Regular check-ins like these will help you spot vulnerabilities before they become full-blown liabilities. Remember, the goal isn't perfection (no business is completely risk-free); it's preparation. With each step you take, you're not just protecting your data—you're building resilience that will serve your business for years to come.

Policies and Procedures: The Unsung Heroes of Cybersecurity

I understand... policies and procedures don't exactly scream excitement. But these unassuming documents are your business's rulebook, the playbook that keeps everything running smoothly and securely.

Think of them as the guardrails on a winding mountain road: not flashy, but absolutely essential.

Why Policies Matter

- **Clarity**: A good policy lays out the ground rules for handling sensitive data, accessing systems, and responding to security incidents. No guessing games—just clear, actionable steps.

- **Accountability**: When employees know what's expected of them, they're less likely to fumble and make mistakes that could lead to costly breaches.

- **Consistency**: A solid set of policies ensures that everyone, from the new hire to the CEO, is on the same page and pulling in the same direction.

What to Include

To make your policies effective, focus on key compliance areas that address real-world risks:

- **Password Requirements**: Spell out expectations for length, complexity, and regular updates. No more "Password123" disasters.

- **Data Protection**: Cover encryption standards, secure storage methods, and guidelines for sharing sensitive information.

- **Incident Reporting**: Provide a clear, simple process for employees to report suspicious activity—who to contact and what steps to take.

- **Acceptable Use**: Set boundaries for how company devices and networks should (and shouldn't) be used, from avoiding risky websites to steering clear of personal downloads.

Keep It Simple

The last thing anyone needs is a 50-page tome that's as impenetrable as a legal textbook. Instead, aim for plain language, pepper in some real-life examples, and make the policy accessible to everyone. Whether it's handed out during onboarding, shared in a company-wide email, or pinned to the breakroom wall, the goal is the same: make it easy to understand and hard to ignore.

With clear, actionable policies in place, you're not just ticking off a compliance box—you're laying the foundation for a business that's secure, accountable, and ready for whatever the digital world throws its way.

Empowering Your Front Line: Compliance Training for Employees

Your employees aren't just part of your cybersecurity defense—they *are* the defense. They're the ones opening emails, handling sensitive data, and navigating potential pitfalls daily. That's why training them on compliance basics isn't just a good idea; it's essential. Think of it as a "compliance bootcamp" for your team, with the goal of turning every employee into a confident, informed defender of your business.

What to Cover

Start with the essentials, the must-knows that can make a real difference:

- **Handling Sensitive Data**: Teach employees the best practices for storing, sharing, and protecting sensitive information, from customer records to internal documents.

- **Spotting Phishing Emails**: Equip them with the skills to identify those too-good-to-be-true offers or urgent, suspicious requests lurking in their inbox.

- **Following Security Policies**: Walk them through your company's specific policies, ensuring they understand not just the "what" but the "why" behind each rule.

Make It Regular

Cyber threats evolve constantly, and so should your training. Treat compliance training as an ongoing process, not a one-and-done event. Schedule sessions quarterly, or at the very least annually, to keep your team sharp. And don't just repeat the same material—update the training to reflect new threats and changing regulations.

Keep It Engaging

Let's be real—no one looks forward to another dull PowerPoint marathon. Make training sessions interactive and memorable:

- Use **quizzes** to test knowledge and add a competitive edge.

- Incorporate **real-world examples** to show how compliance applies to their day-to-day tasks.

- Add **interactive scenarios** or role-playing exercises to practice responses in a risk-free environment.

Why It Matters

A well-trained team isn't just a box ticked off on your compliance checklist—it's your greatest asset. When employees understand the *why* behind compliance efforts, they're more likely to take it seriously, apply it consistently, and even go the extra mile to protect your business.

Because here's the truth: no matter how sophisticated your systems are, it's the people using them who make or break your defenses. With the right training, your team can become a first line of defense that's proactive, informed, and ready for anything.

Working with Third-Party Experts

The truth is that compliance can feel like wading through quicksand while deciphering an ancient code. No one expects you to be an expert in every regulation, acronym, or technical detail, and that's perfectly okay. That's where third-party professionals come into play. Think of them as your compliance superheroes, swooping in to tackle the tricky stuff so you can focus on running your business.

Cybersecurity Consultants

These are your go-to gurus for identifying gaps in your compliance efforts. Whether it's spotting vulnerabilities in your systems or recommending practical fixes, cybersecurity consultants bring a wealth of knowledge to the table. They're especially helpful if you're unsure where to start or need a second opinion on your current security measures.

Legal Advisors

Regulations like GDPR and HIPAA are packed with legal nuances that can make your head spin. A good legal advisor can help translate the jargon into plain English, clarify your specific obligations, and guide you in avoiding those costly mistakes that keep business owners awake at night. Think of them as your compliance interpreter and safety net rolled into one.

IT Professionals

Need a secure network? Unsure how to implement multi-factor authentication or encrypt sensitive data? Enter the IT professionals. Many IT companies specialize in managing compliance requirements and setting up systems that align with regulatory standards. They can also help monitor your systems to ensure ongoing security and compliance.

Why Asking for Help is Smart

Let's face it—no one wants to be the business owner explaining to customers why their data got leaked. Hiring experts might feel like an extra expense, but it's an investment that pays for itself by sparing you fines, reputational damage, and sleepless nights.

If compliance feels like an uphill battle, don't hesitate to call in reinforcements. After all, even the most seasoned adventurers have a guide, and your business deserves the best tools and support to stay on track.

Using Compliance Tools and Software: Turning a Chore into a Breeze

To be honest, compliance has a bit of a reputation for being tedious. But the good news is that technology can help lighten the load. Today's tools are like having a digital assistant who never forgets a deadline, always keeps the paperwork in order, and has an uncanny knack for spotting issues before they arise. Let's look at some of the best tech options to keep your business running smoothly—and legally.

Compliance Management Platforms: Your Virtual Organizer

Imagine a tool that combines templates, reminders, and audits into one easy-to-navigate dashboard. That's what compliance management platforms bring to the table. They help you track what needs to be done, when it needs to be done, and whether it's been done correctly. Think of it as the Swiss Army knife of compliance—it's got a tool for everything, and it keeps you from scrambling at the eleventh hour.

Data Encryption Software: The Digital Vault

Sensitive data is the crown jewel of your business, and encryption is like locking it up in an unbreakable vault. The beauty of data encryption software is that it works automatically, securing your information as it's stored or shared. Whether you're safeguarding customer details, financial records, or proprietary files, encryption ensures that even if a hacker gets in, they won't be able to read the data.

Monitoring Tools: The Watchful Eye

Who's logging into your systems? What are they accessing? When are they doing it? Monitoring tools keep tabs on these activities, helping you maintain tight access controls and meet regulatory requirements. They're like the security cameras of the digital world—always watching, always documenting, and ready to alert you if something doesn't add up.

Simplify, Streamline, Succeed

The best part about these tools? They transform compliance from a dreaded task into a streamlined process. With the right software in place, you're not just checking boxes—you're building a compliance system that works efficiently, minimizes risks, and gives you more time to focus on growing your business.

Embrace the tech. It's there to make your life easier, and in the world of compliance, that's no small thing.

Creating a Compliance Calendar

Consistency is the secret sauce of compliance, and a calendar is your best friend. Here's how to set one up:

- **Monthly Tasks**: Check software updates, review data back-ups, and test incident response plans.

- **Quarterly Tasks**: Conduct risk assessments, review access controls, and provide team training.

- **Annual Tasks**: Update your security policies, review compliance frameworks, and schedule a consultation with a third-party expert (if needed).

A well-maintained compliance calendar ensures nothing slips through the cracks. It also helps you avoid last-minute scrambles when regulators come knocking.

Staying compliant might feel like climbing a mountain with a backpack full of legal jargon, but here's the thing: it's not as daunting as it seems when you have the right tools and mindset. Compliance isn't just about ticking off boxes on a checklist; it's about safeguarding the very foundation of your business—your employees, your customers, and everything you've worked so hard to build.

By creating a solid framework, maintaining consistency, and ensuring your team is well-informed, you can confidently navigate the maze of regulations. It's about more than avoiding fines or penalties; it's about earning trust, building credibility, and securing the future of your business.

And here's the cherry on top: when you know your business is safe, secure, and compliant, you'll finally sleep a little easier at night. Peace of mind, after all, might just be the best ROI of all.

Your Compliance Starter Kit: An Actionable Checklist

After wading through the often-intimidating waters of cybersecurity regulations and best practices, it's time to simplify things. This compliance checklist is your practical roadmap—a clear, actionable plan to ensure your small business stays aligned with the standards that matter most. Think of it as your cybersecurity "starter kit," designed to keep things manageable and, dare I say, stress-free.

Here's what to include:

1. **Identify Applicable Regulations**Determine which regulations apply to your business (e.g., GDPR, HIPAA, PCI DSS, or others) based on your industry, operations, and customer base.

2. **Secure Sensitive Data**

   - Use encryption for sensitive data storage and transmission.

   - Regularly back up your data and store it securely, including offsite backups.

3. **Set Strong Password Policies**

   - Require unique, complex passwords.

   - Implement multi-factor authentication wherever possible.

4. **Train Your Team**

   - Provide regular training on recognizing phishing scams, safeguarding sensitive data, and following security protocols.

- Encourage a "report anything suspicious" culture.

5. Establish Incident Reporting Protocols

- Create a clear process for reporting potential breaches or suspicious activity.

- Assign a dedicated point of contact for cybersecurity concerns.

6. Monitor Access Controls

- Limit access to sensitive data on a need-to-know basis.

- Use monitoring tools to track who accesses your systems and when.

7. Stay Updated

- Regularly review your compliance framework to ensure it reflects any changes in regulations or your business operations.

- Keep software, systems, and tools updated to protect against vulnerabilities.

8. Document Everything

- Maintain records of your compliance efforts, including policies, training sessions, and risk assessments.

- Be ready to demonstrate your commitment to compliance if ever required.

9. **Seek Professional Guidance**

   ○ If compliance feels overwhelming, consult cybersecurity experts, legal advisors, or IT professionals to fill in the gaps.

This checklist isn't just a to-do list; it's your blueprint for creating a secure and resilient business. By taking these steps, you'll not only meet the necessary standards but also build trust with your customers and protect the future of your business. Safe, compliant, and stress-free—now that's a combination worth striving for!

Mastering Compliance Without the Stress

The secret to navigating compliance isn't trying to conquer it all at once—it's breaking it down into bite-sized, manageable tasks. Think of it like assembling a puzzle: start with the corner pieces, and the rest will gradually come together. Here's how to keep the process straightforward and stress-free:

- **Start Small:** Begin with the fundamentals—secure your passwords, conduct basic employee training, and implement simple, clear policies. These small steps create a solid foundation for protecting your business.

- **Delegate Smartly:** Compliance isn't a one-person job. Assign specific tasks—like managing access controls or conducting policy reviews—to team members. Not only does this ensure nothing gets overlooked, but it also makes compliance a team effort.

- **Use Technology to Your Advantage:** Compliance software can be a game-changer. Automate tedious tasks like tracking updates, monitoring data access, or scheduling reminders for policy reviews. Let the tools do the heavy lifting while you focus on the big picture.

- **Stay Consistent:** Think of compliance as a marathon, not a sprint. Regular check-ins, updates, and training sessions will keep you ahead of the curve and help you adapt to new threats or regulations without the last-minute scramble.

By embracing these simple strategies, you can take the overwhelming out of compliance. It's not about achieving perfection overnight; it's about making steady, intentional progress. With each small step, you're not just ticking boxes—you're strengthening your business's security, building trust with your customers, and setting yourself up for long-term resilience.

Wrapping Up: Compliance as a Cornerstone of Trust

As we conclude this chapter, let's take a moment to see compliance for what it truly is: a cornerstone of trust and a shield that protects your business. Sure, meeting regulations helps you avoid fines and legal headaches, but the true value of compliance goes far beyond simply staying out of trouble.

The Value of Compliance Beyond Avoiding Fines

Compliance is more than a checklist or a set of rules to follow—it's a commitment. It's your way of telling customers, employees, and stakeholders: *We care about protecting your data and safeguarding your trust.* When you prioritize compliance, you're not just checking boxes for regulatory bodies—you're making a proactive investment in the integrity and security of your business.

Following data protection laws and implementing security protocols strengthens your foundation. It reduces risks, builds resilience, and creates a safer digital environment for everyone involved. Compliance isn't just a legal obligation—it's a way to future-proof your business.

Compliance as a Competitive Advantage

In today's world, customers are paying closer attention to who they trust with their data. Headlines about data breaches and privacy concerns have made people more cautious. A strong compliance record can be your secret weapon, setting you apart from competitors.

When you demonstrate that you take security and privacy seriously, you build confidence and loyalty. Compliance becomes a selling point—a symbol of your business's trustworthiness and responsibility. And in industries where reputation is everything, that's a competitive edge worth having.

Staying Proactive and Informed

Regulations and cybersecurity threats are constantly changing, making compliance an ongoing process. It's a bit like tending a garden—leave it unattended, and things can spiral into chaos. Staying proactive, reviewing practices regularly, and keeping your team informed ensures you're always ready for what's next, whether it's a new regulation, an emerging threat, or a technological shift.

The good news? Compliance doesn't have to be daunting. Small, consistent steps can go a long way. The key is to keep improving, keep learning, and keep prioritizing the security of your business and its people.

Looking Ahead: Implementing Advanced Security Measures

Now that you've built a strong compliance foundation, it's time to level up. The next chapter will take you deeper into the tools and technologies that can bolster your defenses—firewalls, endpoint protection, and more. These advanced measures will support your compliance efforts and enhance your overall cybersecurity strategy.

Compliance isn't just about staying legal—it's about building strength and trust. Every proactive step you take reinforces your commitment to your customers and your team. Let's take that commitment to the next level together.

CYBERSECURITY ON A BUDGET

THE BALANCING ACT

Introduction and Disclaimer

L et's get right to it: when it comes to cybersecurity, there's no magical, one-size-fits-all solution that works for every business. Protecting your company on a budget often feels like walking a tightrope and if it doesn't, you might not fully grasp the magnitude of the topic. Yes, there are cost-effective strategies you can implement on your own, and they can absolutely help—blocking some of the more basic threats and giving you a foundation to work from.

But here's the hard truth: these DIY measures are not a substitute for comprehensive security. Think of them as putting band-aids on a bigger wound. They might keep a minor scrape from getting infected, but when a real cyberattack comes knocking, those temporary fixes won't hold up.

Cybercriminals are relentless, and their tactics evolve faster than yesterday's headlines. Basic security tools can stop common threats, but they're no match for sophisticated attacks targeting weak points you didn't even know existed. That's why it's essential to understand what you *can* do on your own—and when it's time to call in reinforcements.

In this chapter, we'll explore cost-effective strategies that small businesses can implement right away while also shedding light on the areas where professional help becomes non-negotiable. The goal isn't to scare you; it's to arm you with knowledge so you can make informed decisions about where to invest your resources for maximum protection.

Because here's the thing: cybersecurity doesn't have to break the bank, but it does require a plan. Let's dive in and figure out how to make your budget work for you without leaving the back door wide open.

The Reality of DIY Cybersecurity

Cybersecurity isn't a battle you win once and move on from—it's an ever-evolving game where the rules are constantly changing, and the opposing team never takes a day off. Unfortunately, for small businesses, this often feels like playing defense against an opponent with unlimited resources. Hackers are relentless, and their tools grow more sophisticated by the day.

What you can do on your own—changing passwords, backing up data, and keeping software up to date—is a fantastic place to start. These steps are the equivalent of locking your front door and maybe even installing a peephole. But here's the reality: cybersecurity isn't just about guarding the front door. If the windows, back door, and basement are left wide open, it's only a matter of time before someone gets in.

A DIY approach to cybersecurity will always leave gaps—some you might not even realize are there. And hackers? They're not just capable of exploiting those gaps; they thrive on them. These vulnerabilities aren't just technical oversights; they're potential disasters waiting to happen, with risks that extend far beyond your systems. We're talking about your reputation, customer trust, and, ultimately, your bottom line.

So, while your DIY efforts are commendable—and necessary—they're just the first step. The next challenge is recognizing where your defenses need to be reinforced and how to build a security plan that leaves as few cracks as possible. Because in this game, you can't afford to let your guard down—not even for a moment.

Stopgap Measures, Not Silver Bullets

The strategies in this chapter are designed with one goal in mind: to help you tackle the cybersecurity essentials without sending your budget into a tailspin. They're practical, affordable, and when applied consistently, can genuinely make a difference. But here's the catch—they're not a substitute for professional-grade cybersecurity measures. Think of them as temporary reinforcements, the kind of defenses that might buy you some time but won't stand up to a determined attacker.

It's also worth addressing the elephant in the room: relying solely on low-cost solutions can create a dangerous illusion of safety. You might feel like you're checking all the boxes, but the reality is, vulnerabilities will persist. Hackers love a false sense of security—it makes their job easier.

That's why the focus of this chapter is to give you tools and techniques that can reduce your risk and provide a strong starting point. But as you move forward, it's crucial to understand that true, ironclad security requires investing in professional resources. Consider this chapter your foundation, not the finished fortress. With these basics in place, you'll be better equipped to identify where to fortify your defenses next.

Cybersecurity: The Insurance You Hope You Never Need

When you stop and think about it, cybersecurity is a lot like insurance. It's not exactly the most thrilling line item in your budget, and it's tempting to skip over or cut corners. But when something does go wrong—and in today's digital world, that's more a matter of when than if—you'll be profoundly grateful for every penny you invested.

Here's the reality: every dollar you save by skimping on professional-grade defenses could cost you tenfold—or more—if your business falls victim to a cyberattack. The fallout isn't just about replacing a few systems or recovering data. It's about lost revenue, legal fees, fines, and the kind of reputational damage that takes years to rebuild, if it can be rebuilt at all.

This chapter is about meeting you where you are—providing affordable, practical steps to shore up your defenses. But let's be clear: these measures are stopgaps, not solutions. They're meant to help you weather the storm until you're in a position to build the comprehensive cybersecurity strategy your business truly needs. Think of this as a foundation—something to work with while you plan for the bigger, more protective structure that will safeguard your business for the long haul.

Let's get straight to it: there are practical steps you can take right now to bolster your cybersecurity without emptying your wallet. These aren't flashy, high-tech solutions, but they are effective starting points. Think of them as the training wheels for your business's security journey—useful for keeping you upright, but not quite enough to tackle the big hills on your route.

The tips and tools in this chapter are designed to help you stay in the game, to buy you time, and to reduce your risk while you work on building a long-term plan. But remember, these measures are just the beginning. Cybersecurity isn't a sprint; it's a marathon. To cross the finish line, you'll need a robust strategy, professional support, and a commitment to ongoing improvement.

So, grab your notepad (or your favorite digital tool), and let's dive into practical steps you can implement today. Each one is a step closer to protecting your business—but don't lose sight of the bigger picture. This is about setting the stage for something stronger, something smarter, and something that lasts.

Cybersecurity on a Shoestring: Starting Small

Let's be honest—cybersecurity can seem like an exclusive club for Fortune 500 companies with their deep pockets and armies of IT professionals. For small businesses, it's easy to feel like you're left to fend for yourself. But here's the silver lining: you don't need a sprawling budget to start protecting your business.

The key is to focus on the essentials—steps that make a real impact without costing a fortune. While these budget-friendly strategies won't replace the ironclad security systems of larger enterprises, they can help reduce your risks and buy you valuable time. Think of them as the cybersecurity equivalent of locking your doors and windows—not foolproof, but far better than leaving everything wide open.

So, if you're ready to roll up your sleeves and get practical, let's explore how to make the most of what you've got, one step at a time. Even on a shoestring budget, a little effort can go a long way toward keeping your business safer.

Starting with What You Have

Here's the thing—you don't have to be decked out with the latest tech gadgets or employ a 24/7 cybersecurity team to start making a difference. Small, actionable measures can go a long way in protecting your business, and some of them won't cost you a dime.

Take strong passwords, for instance. They're not glamorous, but they work. Pair that with enabling multi-factor authentication (MFA) and training your team to spot phishing scams, and you've already built a pretty sturdy first line of defense. These steps might not fend off a sophisticated hacker, but they'll make your business a less attractive target for the more opportunistic cybercriminals out there. Pair these with a password management tool, as suggested earlier in this book, to effortlessly create and manage sophisticated, complex passwords. These tools take the stress out of keeping track of multiple passwords, allowing you to maintain strong security without the hassle.

Budget-friendly cybersecurity is all about doing what you can, with what you've got, right now. It's not about perfection—it's about taking proactive steps to reduce your risks while laying the groundwork for more robust defenses in the future. Because let's face it, hoping for the best isn't much of a strategy. Even small, consistent efforts can send hackers looking for an easier target.

Small Businesses: The Surprising Favorite of Hackers

It's easy to think that cybercriminals are only after big fish—those massive corporations with deep pockets and mountains of data. But here's the cold, hard truth: small businesses are prime targets. Why? Because hackers know smaller companies often lack the sophisticated defenses of their larger counterparts. To a cybercriminal, your small business might as well have a sign on the door that says, "Easy pickings."

If you're working with a limited budget and assuming that makes you an unlikely target, it's time for a reality check. Attackers don't care about your size; they care about your vulnerabilities.

Prioritizing Cybersecurity Without Breaking the Bank

The good news is that protecting your business doesn't require a six-figure IT budget to get started. Even modest, cost-effective measures can make a significant difference. Ignoring cybersecurity altogether, though? That's a gamble you can't afford to take.

Think of it this way: a single breach could cost your business far more than investing in a few basic preventative steps today. And while professional-grade solutions might not be feasible right now, the measures you can afford to implement today will reduce your immediate risks and buy you time to plan for more robust defenses down the road.

Because in the end, the question isn't whether your business will be targeted—it's whether you'll be ready when it happens.

Cybersecurity: A Practical Investment, Not an Optional Expense

Think of cybersecurity as a necessary investment, not a luxury. It's no more optional than locking your front door at night. Sure, a state-of-the-art alarm system might be out of reach for now, but that doesn't mean you'd leave the door wide open. Instead, you'd start with affordable basics—a sturdy lock, a motion-sensor light—things that make intruders think twice. Cybersecurity is no different.

The key is shifting your mindset. Instead of seeing cybersecurity as a massive, unattainable expense, view it as a process—one that you build step by step. Each small action, like enforcing a strong password policy, training your team to spot phishing emails, or keeping software up to date, is part of that process.

These small measures might not seem like much on their own, but layered together, they create a stronger, more resilient defense. It's not about perfection or going from zero to fortress overnight. It's about doing what you can today, knowing that every step strengthens your business and protects what matters most.

Laying the Foundation for Cybersecurity

The strategies in this chapter are all about getting started on a budget—taking those first critical steps toward safeguarding your business. Think of them as the foundation for a larger, more comprehensive approach to cybersecurity. While these measures won't cover every angle, they're an excellent starting point, especially if resources are limited.

Let's dive into how you can make the most of what's available, leverage cost-effective solutions, and take meaningful steps to protect your business—all without straining your finances. Remember, these strategies are not the finish line; they're the beginning of a journey toward stronger, more resilient defenses.

High-Impact, Low-Cost Security Tips

Cybersecurity doesn't have to begin with a massive financial outlay. In some cases, there are simple, low-cost actions that pack a punch when it comes to protecting your business. These straightforward measures act as the sturdy foundation upon which you can build a more comprehensive security framework as your resources grow.

Think of it like patching a leaky roof—small, targeted fixes can prevent a flood while you save up for a full renovation. The key is to start where you are, using affordable tools and common-sense practices to minimize your risks today while planning for tomorrow's upgrades.

Here's how you can boost your defenses without breaking the bank. Many of the following items have already been covered (yes, I know I'm repeating myself) but I want to reinforce the basic steps to help you do as much as you can on your own, if that's what you choose.

Empowering Your Employees: The Best Budget-Friendly Defense

Your employees are not just part of your business—they're your front-line defenders against cyber threats. However, without the proper training, they can also become your Achilles' heel. The good news? Teaching your team to recognize common red flags, like phishing emails or suspicious links, doesn't have to cost a fortune.

Start small and keep it simple. Quick, informal training sessions can be surprisingly effective. Show your team real-world examples of phishing emails, pointing out the telltale signs—dodgy grammar, odd sender addresses, or over-the-top urgency. Then, make it clear:

no matter how legitimate a request might seem, unusual asks should always be verified, even if they appear to come from someone familiar.

You can also spice things up with free online resources and quizzes. Turn learning into a game—employees will engage more, retain more, and maybe even enjoy themselves along the way. And remember, a little education can go a long way toward preventing costly mistakes. An informed team isn't just a safeguard; it's a cost-effective shield for your entire business.

Keep It Current: The Power of Software Updates and Patch Management

Software updates may not be the first thing on your mind as you start your day, and I don't expect them to be. But they can't be ignored either! They pop up at the most inconvenient times, and it's all too tempting to hit "Remind Me Later" and move on. But here's the thing: skipping updates is like leaving your front door wide open with a neon sign that says, "Welcome, hackers!" Cybercriminals thrive on finding vulnerabilities in outdated software, and those updates? They're the locks and deadbolts that keep them out.

The best part about this defense? It's completely free and requires minimal effort. Most software updates cost nothing but a few clicks (or better yet, no clicks at all if you enable automatic updates). Whether it's your operating system, antivirus software, or the apps you rely on daily, keeping everything up to date ensures that your digital doors are securely locked.

Think of updates as maintenance for your digital world—quick, effective, and essential. Sure, they may interrupt your workflow for a few minutes, but that's a small price to pay compared to the time and money a cyberattack could cost you.

Multi-Factor Authentication: The Digital Deadbolt

If a password is your digital lock, then Multi-Factor Authentication (MFA) is the sturdy deadbolt that keeps intruders out. Even if a hacker somehow gets their hands on your password, MFA adds an extra layer of protection—usually in the form of a code sent to your phone or generated by an app. This simple additional step transforms your account security from a "one and done" defense into a multi-layered fortress.

The beauty of MFA is its accessibility. Most platforms—think email, banking apps, and cloud storage services—offer MFA for free. Setting it up usually takes just a few minutes, and the payoff is huge. It's one of the most straightforward and cost-effective ways to keep cybercriminals at bay.

So, next time you're asked to enable MFA, don't think of it as an extra hassle—think of it as installing a deadbolt on your digital front door. It's a small step that makes a big difference.

Strong Passwords: Your First Line of Defense

Let's face it—weak passwords like "123456" or the ever-popular "password1" are about as effective as a screen door on a submarine. Cybercriminals love them because they're easy to guess, giving hackers a direct line into your accounts.

Encourage your team to think creatively when crafting passwords. A strong password is a mix of uppercase and lowercase letters, numbers, and symbols—and it avoids personal details like names or birthdays, which are easy for attackers to figure out. Think of it as creating a secret code only you and your systems understand.

Worried about remembering all those complicated passwords? That's where password managers come in. These handy tools securely store and even generate strong, unique passwords for all your accounts. Many offer free versions, saving you time, brainpower, and the temptation to reuse "Fluffy123" across every account.

Strong passwords may seem like a small step, but they're a giant leap in keeping your accounts and business safe. After all, your cybersecurity is only as strong as the weakest password in your arsenal.

Data Backups: Your Safety Net When Things Go Wrong

Imagine losing all your business data in one fell swoop—a ransomware attack locks you out, a hardware failure wipes your files, or an accidental keystroke sends critical documents into the void. It's the stuff of nightmares. That's where data backups swoop in as your safety net, ensuring those precious files aren't gone forever.

The good news? A solid backup strategy doesn't have to break the bank. Many cloud storage services offer free or budget-friendly plans that automatically save your files, so you don't have to think about it. Pair that with an external hard drive—a one-time purchase—and you've got a double layer of protection.

Here's the golden rule: don't put all your eggs in one basket. Use both cloud storage and physical backups for maximum security. The cloud provides convenience and offsite protection, while external drives give you control over a secure offline copy.

And here's a pro tip: don't just set it and forget it. Test your backups periodically to make sure they actually work when you need them. After all, a backup that can't be restored is about as useful as a parachute with holes. By taking these simple steps, you can breathe easier knowing your data is safe, no matter what happens.

Access Controls: The "Need-to-Know" Approach

Not everyone on your team needs the keys to every digital door. That's the essence of the "principle of least privilege"—a fancy way of saying that employees should only have access to the data and systems they need to do their jobs, and nothing more.

Think about it: does your marketing team really need to poke around in payroll files? Or should your receptionist have admin-level control over sensitive systems? Probably not. Limiting access not only keeps your data safer but also reduces the chance of accidental mishaps or internal security breaches.

The best part? Most systems already allow you to set up access controls for free. It's as simple as assigning permissions based on roles or tasks. Make it a habit to review these permissions regularly, especially as employees move to new roles or take on additional responsibilities.

By implementing thoughtful access controls, you're not just protecting sensitive data—you're creating a more streamlined, secure work environment where everyone has exactly what they need to succeed, and nothing they don't.

Small Steps, Big Impact

These tips go beyond saving you money and even beyond protecting your data. They lay the foundation to actually embed cybersecurity into the DNA of your business. By weaving these budget-friendly measures into your daily operations, you're creating a culture where security isn't an afterthought—it's second nature.

Each step you take, no matter how small, contributes to a stronger defense. Encouraging employees to think critically about emails, setting up multi-factor authentication, or running regular backups might seem like minor actions, but collectively, they can create significant barriers for cybercriminals.

Cybersecurity isn't just about expensive tools and technologies; it's about fostering awareness, staying vigilant, and doing the best with what you have. Every step forward—even on a limited budget—is a step toward making your business a less appealing target for attackers. Remember, it's not about perfection; it's about progress.

Cost-Conscious Cybersecurity: Maximizing Impact on a Budget

The great news is that robust cybersecurity doesn't always require a hefty price tag. With the right tools—and a bit of strategic thinking—you can set up defenses that pack a punch without emptying your wallet. Many affordable, and even free, options are available to help you protect your business effectively.

From free antivirus software to simple data backup solutions, these tools offer practical, budget-friendly ways to reduce risks and fortify your digital environment. Let's dive into some smart, cost-conscious options that deliver maximum security for minimal investment.

Antivirus and Anti-Malware Software: Your First Line of Defense

You don't need a blockbuster budget to keep viruses and malware at bay. Plenty of antivirus programs offer free versions that deliver reliable, no-frills protection for small businesses. Trusted names like Avast, Bitdefender, and Windows Defender provide solid baseline security to safeguard your systems.

Take Windows Defender, for example. Built directly into most Windows operating systems, it's evolved into a surprisingly effective tool, offering robust, real-time protection against many common threats—all at no extra cost.

Free versions are a fantastic starting point, especially for businesses with simple security needs. And when you're ready to take your defenses to the next level, many of these programs offer affordable premium tiers with perks like advanced threat detection, ransomware protection, and round-the-clock support. Whether you stick with the free plan or upgrade down the line, these tools can provide peace of mind without breaking the bank.

Again, I do want to caution you that not all of these solutions are created equal and things do change. I'm not going to mention names but over the years, we've found that yesterday's Trusted Antivirus (AV) solution, is no longer performing up to par with the newer cyber threats. One of the things that the cybersecurity specialists do is to stay on top of the threats and the best ways to protect against them, which often means using a different AV solution. In addition to that, many of these specialists have around-the-clock monitoring in case something does make it through.

Firewalls: Your Digital Gatekeepers

Think of firewalls as the security guards of your digital world, keeping an eye on who's coming and going and deciding whether they should be allowed through. The good news? You probably already have one. Many routers come with built-in firewalls, and both Windows and macOS operating systems include them as well. While these tools are limited, they are far better than nothing at all.

Getting started is straightforward: check that your firewall is enabled and properly configured. This one step can significantly reduce your exposure to online threats. If you're unsure about the settings, a quick consultation with your IT provider can give you peace of mind.

Firewalls might not be flashy, but they're one of the most reliable ways to block unwanted traffic and keep your network secure. With just a little effort, you can put a strong first line of defense between your business and potential cyber threats.

Password Management Tools: Ditch the Spreadsheet

Struggling to remember all your passwords or relying on the timeless (but oh-so-risky) sticky-note-on-the-monitor method? Let's level up your password game. Password management tools securely store and organize your credentials, ensuring they're strong, unique, and easily accessible when you need them.

Affordable or even free options like LastPass, Bitwarden, and KeePass are great starting points. LastPass and Bitwarden offer free tiers ideal for small businesses, featuring secure password storage and convenient autofill capabilities. For the more tech-savvy or DIY-inclined, KeePass is a completely free, open-source tool that's highly customizable.

Whichever option you choose, these tools save time, reduce frustration, and, most importantly, boost security. With a good password manager, you can finally ditch the sticky notes and sleep easier knowing your accounts are well-protected.

Data Encryption: Locking Your Digital Door

Think of data encryption as a magical way to turn your information into gibberish for anyone without the right key. Even if hackers intercept your data, encryption ensures it's utterly useless to unauthorized eyes—like handing them a locked diary with no key in sight.

The best part? You're probably already using some form of encryption without realizing it. Many cloud storage providers, like Google Drive and OneDrive, have built-in encryption baked into their services. For those handling especially sensitive information or craving more control, free tools like VeraCrypt offer a straightforward way to encrypt files right on your devices.

Despite how it sounds, encryption isn't reserved for tech wizards. These tools make it simple, affordable, and highly effective. It's like putting your data in a virtual safe, ensuring that even if someone finds it, they won't have the combination to open it.

Using Free Versions of Paid Security Software: Start Small, Scale Up

Many premium security tools offer free versions or trials that can deliver robust basic protection while you assess whether they're worth the investment. For example:

- **Malwarebytes** provides a free version that's excellent for on-demand malware scanning.

- **ZoneAlarm** offers a free firewall and antivirus combination.

- **Sophos Home** has a free tier that protects up to three devices with real-time malware protection.

These free versions often lack advanced features like enterprise-level reporting, but they can serve as a decent starting point while your business grows.

Affordable Security Boosters: Plugins That Work Behind the Scenes

If your business has a website or does email communications—come on, you know you do—adding budget-friendly plugins can be a game-changer for your cybersecurity. These tools quietly do the heavy lifting, providing an extra layer of protection without disrupting your daily operations or draining your wallet.

Take Wordfence, for instance. It's a popular plugin for WordPress sites that offers firewall protection, malware scanning, and login security—all in its free version, which is perfect for smaller websites. It's like hiring a 24/7 security guard for your site who doesn't demand coffee breaks.

For email protection, browser extensions like anti-phishing tools or services like Avanan can block suspicious messages and malicious links before they ever land in your inbox. These tools add a valuable safety net, catching threats early and reducing the risk of someone accidentally clicking on something they shouldn't.

The beauty of these plugins is their simplicity. They work quietly in the background, requiring minimal maintenance while adding maximum protection. It's an affordable way to upgrade your defenses and give you peace of mind without the need for a hefty investment.

While they are not without their complexities, they work well. They will require policy tweaking to work as intended, but they will still protect you out-of-the-box.

Building a Strong Foundation Without Breaking the Bank

These low-cost and free tools are a reminder that strong cybersecurity doesn't have to break the bank. By tapping into solutions like firewalls, password managers, and email security plugins, you can tackle many of the vulnerabilities small businesses face without a hefty investment.

But here's the thing—these tools are just one piece of the puzzle. To get the most out of them, you'll need to combine them with consistent maintenance, ongoing employee training, and a proactive approach to staying secure. It's like owning a great lock for your front door; it's only effective if you actually use it and check that it's working properly.

With these tools in your cybersecurity toolkit, you'll be better prepared to protect your business, fend off common threats, and keep costs manageable. While they're a fantastic starting point, think of them as stepping stones on the path to a comprehensive security strategy—one that grows with your business and continues to safeguard what you've worked so hard to build.

Affordable Tools: The Building Blocks

These low-cost and free tools prove that securing your business doesn't have to empty your wallet. Firewalls, password managers, and email security plugins provide practical, effective defenses against many of the threats small businesses face—all without requiring a major financial commitment.

However, here's the catch: tools alone aren't enough. They're only as good as the habits and practices you build around them. Think of them like the seatbelt in your car—essential for safety, but only effective if you actually buckle it every time you drive. Regular up-dates, employee training, and a mindset of vigilance are the real keys to maximizing their value.

By incorporating these tools into your cybersecurity strategy, you're laying the foundation for a more secure operation. They're your first line of defense—practical, affordable, and effective—but they're also just the beginning. Use them as a springboard for creating a robust security framework that grows alongside your business, pro-tecting everything you've worked so hard to achieve.

Cybersecurity Tasks You Can Tackle Yourself

You don't need a PhD in IT or a wall of blinking servers to make significant strides in cybersecurity. In fact, some of the most effective defenses are entirely within reach for non-techies. With a little research and some dedication, you can take on these essential tasks:

- **Managing Software Updates:** Keeping your devices and programs current is like locking the windows before you leave the house. Updates often patch vulnerabilities that cy-bercriminals are itching to exploit. Most modern systems make this easy with automated updates—just make sure they're turned on and running.

- **Enabling Multi-Factor Authentication (MFA):** Adding MFA to your accounts is like bolting the door after locking it. It's quick to set up, and most platforms walk you through the process step by step. By requiring a second form of verifica-tion, like a text code or app confirmation, MFA dramatically reduces the chances of unauthorized access.

- **Basic Employee Training:** Teaching your team to spot

phishing emails, create strong passwords, and follow simple cybersecurity rules doesn't require a professional trainer.
Regular reminders, clear examples, and an open-door policy
for questions can go a long way in building awareness and
confidence.

These tasks are perfect for small businesses operating on tight budgets. They're simple, manageable, and—most importantly—effective.
With a proactive approach and a willingness to roll up your sleeves,
you can create a solid foundation of security that makes a real difference.

When DIY Isn't Enough

As empowering as it is to tackle certain cybersecurity tasks on your
own, there are limits to how far a DIY approach can take you. Cybersecurity is a constantly shifting battlefield, and some areas demand
expertise that goes well beyond the basics. For instance:

- **Vulnerability Assessments:** Pinpointing weak spots in
  your network isn't just about spotting outdated software or
  weak passwords. It requires an in-depth understanding of
  technical systems and the tactics hackers use to exploit them.

- **Penetration Testing:** Think of this as hiring a professional "white-hat" hacker to test your defenses. Simulating
  a real-world attack to see how your system holds up isn't
  something to dabble in—it's a job for trained professionals
  who know how to push your security to its limits without
  causing harm.

- **Advanced Security Systems:** Setting up complex protections like firewalls, intrusion detection systems, or endpoint security requires precise configurations. These tools are only as effective as the person implementing them, and mistakes can leave critical vulnerabilities wide open.

The Reality Check

While DIY is a fantastic way to lay a solid foundation, there are moments where going it alone isn't just ineffective—it's risky. When the stakes are high, trial and error isn't a luxury you can afford. Knowing when to call in experts is just as important as knowing when you can handle things yourself. It's about striking the right balance between self-reliance and professional support to build a truly secure business.

Knowing When to Call in the Experts

There's no shame in admitting when a job is beyond your skill set—especially when it comes to cybersecurity. There comes a point when the risks of going it alone far outweigh the costs of professional help. Here are a few scenarios where bringing in a cybersecurity expert or a managed service provider (MSP) isn't just a good idea—it's essential:

- **You Handle Sensitive Customer Data:** If your business processes credit card transactions, manages personal health information, or stores sensitive customer details, you're holding a goldmine of valuable data that hackers are eager to exploit. Protecting this data isn't just smart—it's a legal and ethical responsibility.

- **You Need to Meet Compliance Standards:** Regulations like GDPR, HIPAA, or PCI DSS aren't optional, and they come with strict requirements that can be overwhelming to navigate on your own. An expert can help you understand what's needed, implement the proper safeguards, and keep your business on the right side of the law.

- **You've Experienced a Previous Security Breach:** If you've already been through the nightmare of a cyberattack, you know how costly and stressful it can be. Hiring a professional can help shore up your defenses, prevent future incidents, and restore confidence—both for you and your customers.

The Investment That Pays Off

Bringing in outside help might feel like a big expense upfront, but think of it as insurance. It's a proactive step to protect your business, your customers, and your reputation from risks that could cost far more in the long run. Sometimes, the best way to safeguard what you've built is to call in reinforcements.

Affordable Ways to Access Professional Cybersecurity Expertise

Hiring professional cybersecurity services doesn't have to mean draining your budget or committing to long-term contracts. There are cost-effective options that allow you to get the expertise you need without breaking the bank:

- **Project-Based Consultants:** Need help with a specific task? Instead of hiring a full-time cybersecurity expert, consider bringing in a consultant for a defined project. Whether it's a one-time vulnerability assessment, penetration testing, or setting up your security infrastructure, a consultant can tackle the job efficiently and effectively—without adding a permanent salary to your payroll.

- **Subscription-Based MSPs:** Managed service providers (MSPs) offer ongoing cybersecurity monitoring and support for a predictable monthly fee. These services often include essentials like firewall management, threat detection, and regular updates, making them a great choice for small businesses. Many MSPs specialize in working with smaller organizations and understand the need for robust security on a tight budget.

Professional Help, Budget-Friendly Approach

By focusing on targeted help—whether it's for a single project or through a monthly subscription—you can strike a balance between expert protection and affordable spending. These options give you access to professional-grade cybersecurity without the financial strain of hiring an in-house team, so you can safeguard your business without sacrificing your bottom line.

Striking the Right Balance

DIY cybersecurity is a fantastic starting point, especially for small businesses with limited resources. But as your business grows—or the stakes get higher—there will come a time when professional help is not just an option, but a necessity.

By recognizing your limits, you can focus on what you do best while leaving complex, high-stakes tasks to the experts. Prioritizing critical areas like vulnerability assessments, compliance, and advanced security systems ensures your efforts are focused where they matter most.

Seeking professional assistance doesn't have to mean breaking the bank, either. With options like project-based consultants and subscription-based managed service providers (MSPs), you can access expert guidance in a way that aligns with your budget.

Peace of Mind is the Real ROI

At the end of the day, the goal of a smart cybersecurity strategy isn't just about protection—it's about peace of mind. Knowing that you've taken the right steps, sought the right help, and prioritized your business's safety allows you to focus on what matters most: growing your business and serving your customers.

Remember, cybersecurity is an ongoing process. By balancing DIY efforts with professional expertise when it counts, you're setting up your business to stay one step ahead of the ever-evolving threats of the digital world.

Putting It All Together: Your Budget-Friendly Cybersecurity Plan

You've explored practical, low-cost ways to safeguard your business, and now it's time to put those insights into action. The idea here isn't to do everything at once or aim for perfection—it's to create a manageable, step-by-step plan that fits your budget and grows with your business. Let's break it down.

Creating a Cybersecurity Action Plan

Start small and focus on what you can realistically accomplish with the resources you have today. Here's a sample roadmap to get you started:

1. **Train Your Team**: Use free online resources to teach employees the basics of cybersecurity, like spotting phishing scams and using strong passwords.

2. **Turn On MFA**: Enable multifactor authentication (MFA) on critical accounts to add an extra layer of security.

3. **Review Your Basics**: Check that all software is up-to-date, passwords are strong, and firewalls are enabled.

This starter plan is flexible, giving you room to build on these initial steps as your needs—and your budget—grow.

Focus on High-Impact, Low-Cost Steps

You don't need to overhaul your entire system to see meaningful improvements. Start with the measures that offer the most protection for the least cost:

- **Employee Training**: A quick refresher on spotting suspicious emails and links can prevent costly mistakes.

- **Software Updates**: Regular updates patch vulnerabilities that hackers love to exploit.

- **Password Practices**: Strong, unique passwords (managed with free or low-cost password managers) go a long way toward securing your accounts.

By focusing on these high-priority areas, you'll see immediate benefits without stretching your finances.

Making Cybersecurity Routine

Think of cybersecurity as an ongoing habit rather than a one-time effort. Set up a schedule for essential tasks, like:

- **Monthly Software Updates**: Dedicate time each month to review and update your tools.

- **Quarterly Training Refreshers**: Keep your team sharp with short, interactive sessions on cybersecurity basics.

- **Regular Backups**: Ensure that critical data is backed up and easily retrievable.

By integrating these small tasks into your operations, you'll create a rhythm that keeps your defenses strong without overwhelming your team.

Tracking Your Progress

Your cybersecurity plan isn't set in stone—it should evolve as your business grows and threats change. Keep a record of what you've done, like enabling MFA or conducting employee training, and note any areas that need more work.

Every few months, evaluate your progress:

- Are your employees better at identifying phishing scams?

- Is updating software and running backups part of your routine?

Celebrate those wins, but also be ready to adjust. As your business scales, your needs will change, and your cybersecurity strategy should keep pace.

Building for the Future

While this chapter focuses on low-cost solutions, remember that they're just the starting point. These steps give you the breathing room to plan for more comprehensive protections down the line. Cybersecurity isn't just about checking boxes—it's about protecting your hard work, your team, and your customers.

By starting small, staying consistent, and revisiting your plan regularly, you'll build a resilient foundation that grows with your business. It's not just about saving money—it's about creating peace of mind in an unpredictable digital world.

Wrapping Up: The Power of Small Steps in Cybersecurity

Cybersecurity doesn't have to be an overwhelming or costly endeavor. In fact, the strongest defenses often begin with the simplest actions. Training your team to spot red flags, keeping your software updated, and enabling multifactor authentication are straightforward measures that pack a powerful punch in protecting your business.

Why Simple Measures Work

The secret to effective cybersecurity isn't a sprawling budget or a team of tech wizards—it's consistency. Every small, thoughtful step you take builds on the last, creating a cumulative effect that significantly reduces your vulnerabilities. From running backups to reviewing access controls, these manageable actions create a safety net that protects your business from common threats.

Staying Proactive: You've Got This

Cybersecurity isn't about tackling everything all at once; it's about taking steady, intentional steps forward. Use the tools and resources available to you, prioritize what matters most, and remember that every effort counts. Each action, no matter how small, brings you closer to a secure, resilient business. You've already laid the groundwork—keep building on it.

Looking Ahead

With the basics covered, it's time to think bigger. In the next chapter, we'll explore advanced security solutions for businesses ready to elevate their defenses. From cutting-edge tools to expert strategies, we'll dive into what it takes to go beyond the fundamentals and create a comprehensive cybersecurity strategy. Stay tuned—your business's digital safety is about to reach new heights.

AI-POWERED DEFENSES

USING TECHNOLOGY TO STAY AHEAD

Introduction to AI in Cybersecurity

In today's world, where cyber threats evolve faster than we can hit "update," our defenses need to be just as swift and adaptable. Enter artificial intelligence (AI)—a technology shrouded in intrigue but undeniably reshaping how businesses tackle cybersecurity. Whether you're running a tech empire or a small local shop, AI is no longer a luxury; it's becoming an essential part of staying secure in the digital age.

What is AI, and How Does it Apply to Cybersecurity?

At its heart, artificial intelligence is about teaching machines to think, learn, and adapt—just like humans, but at lightning speed and without the need for coffee breaks. In cybersecurity, AI acts as your digital bodyguard, tirelessly analyzing patterns, detecting threats, and even responding to attacks faster than any human ever could.

Picture this: AI spots a phishing attempt not by recognizing a specific scam but by understanding subtle deviations in email patterns. It identifies malware, not because it's seen it before, but because it recognizes its behavior. AI doesn't just guard against yesterday's threats—it learns and adapts to anticipate the tactics of tomorrow's cybercriminals.

Think of it as the Sherlock Holmes of cybersecurity—sharp-eyed, always ahead of the game, and remarkably good at uncovering schemes before they unfold.

Why AI is Not Just for Big Corporations

If the mention of AI brings to mind sprawling tech headquarters and billionaire budgets, it's time for a reality check. Thanks to advancements in technology, AI-powered tools are becoming more accessible and affordable—even for small businesses.

These tools can monitor network activity, flag suspicious behavior, and strengthen password security, all without requiring a team of IT experts. Platforms tailored for small businesses put the power of AI within reach, leveling the playing field against cyber threats that were once the bane of only the largest enterprises.

So, if you've been thinking, *"AI sounds great, but isn't it just for the Amazons and Googles of the world?"*—think again. The same technology safeguarding global corporations is now your ally, ready to help your business stay one step ahead.

In the next section, we'll explore how small businesses can integrate AI-powered defenses into their operations without breaking the bank or requiring a Ph.D. in computer science. Let's see what this game-changing technology can do for you.

How AI Levels the Playing Field for Small Businesses

Here's the thing about AI: it's not just about adding fancy tech to your toolbox—it's about giving small businesses the edge they need in a world where cyber threats are increasingly complex. Let's break down why AI is a game-changer for small operations:

Enhanced Threat Detection

Imagine sifting through mountains of data to spot subtle, suspicious patterns—it's the cybersecurity equivalent of finding a needle in a haystack. AI can process these massive amounts of data in real-time, identifying anomalies that would take human analysts hours—or even days—to catch. From unusual login attempts to subtle phishing schemes, AI has the eagle eyes your business needs to detect threats before they cause damage.

Faster Response Times

When it comes to cyberattacks, every second counts. The beauty of AI is its ability to act immediately. It doesn't stop to deliberate or wait for instructions; it shuts down threats as soon as they're detected, preventing an issue from spiraling into a full-blown crisis. Think of it as having a 24/7 security guard who never takes a coffee break.

Efficient Resource Management

Let's face it—small businesses rarely have the luxury of a dedicated IT team monitoring systems around the clock. AI takes on the heavy lifting, from analyzing logs to responding to minor security incidents, freeing up your team to focus on what matters most: growing your business.

Why AI Matters Now More Than Ever

By integrating AI into your cybersecurity strategy, you're no longer playing defense against cybercriminals—you're actively staying ahead of them. It's the great equalizer, giving small businesses access to cutting-edge protection that was once only available to industry giants.

AI isn't just a glimpse into the future; it's a practical, affordable reality for today's businesses. Whether you're just starting out or well-established, AI-powered defenses give you the tools to safeguard what you've built and keep bad actors at bay.

In the sections ahead, we'll explore actionable ways to integrate AI into your cybersecurity plan. From selecting the right tools to maximizing their potential, we'll show you how to harness this transformative technology without blowing your budget. Let's dive in and see what AI can do for you.

How AI Can Help Small Businesses

For small businesses, staying ahead of cyber threats can feel like an uphill battle. Limited resources, lack of dedicated IT staff, and an ever-changing threat landscape make cybersecurity a daunting task. Enter AI-powered tools—designed to level the playing field and give small businesses the ability to defend themselves with efficiency and precision. Let's dive into how AI can transform your approach to cybersecurity.

AI: The Tireless Watchdog of Cybersecurity

Imagine standing at the shore of an ocean, tasked with finding a single drop of oil in the vast expanse of water. For a human, it's an overwhelming and practically impossible job. But for AI, it's a piece of cake.

AI-driven tools are built for this kind of challenge. They scan mountains of data in seconds, zeroing in on anomalies and patterns that even seasoned cybersecurity professionals might miss after hours—or even days—of analysis. AI doesn't just find the needle in the haystack; it sifts through the hay, categorizes it, and identifies any potential threats before they've had a chance to become real problems.

For small businesses without the luxury of a full-time cybersecurity team, these tools are nothing short of revolutionary. They work tirelessly, 24/7, serving as a vigilant watchdog that never blinks. AI ensures that threats are detected and addressed in real-time, offering a level of protection that was once out of reach for businesses without massive budgets.

In a world where time and resources are often stretched thin, AI becomes your greatest ally—relentlessly monitoring your digital landscape so you can focus on running your business with confidence.

Anomaly Detection: AI's Built-In Suspicion Radar

One of the most impressive tricks AI brings to the cybersecurity world is its knack for spotting anomalies. Imagine having a security system so attuned to your business that it knows exactly what "normal" looks like—when files are usually accessed, which devices log in, and what times emails typically go out. Over time, AI learns your network's unique rhythms, almost like it's memorizing your business's digital heartbeat.

Then, when something unusual happens—a login attempt from an odd location, a surge of unexpected data transfers, or a system accessed at 3 a.m.—AI immediately raises the alarm. It's like having a guard dog that doesn't just bark at everything but instead knows exactly when something's truly out of place.

This proactive approach is a game-changer. Instead of scrambling to clean up the aftermath of a cyberattack, you're catching potential threats in their infancy, stopping them before they grow into full-blown crises. AI doesn't wait for trouble—it detects, analyzes, and acts while the rest of us are still sipping our morning coffee.

AI-Powered Tools: Your Digital Bodyguard

AI-powered tools are like having a hyper-vigilant bodyguard watching over your emails, downloads, and website interactions 24/7. Imagine someone tirelessly scanning every incoming file, scrutinizing every email, and analyzing every link for even the faintest whiff of danger. That's precisely what AI does—only it does it faster and more accurately than any human ever could.

Let's say a phishing email tries to worm its way into your inbox, complete with an attachment carrying ransomware. Before you even have a chance to open it, the AI tool has already spotted the threat. It isolates the suspicious file, blocks it from spreading, and neutralizes the risk—all in the blink of an eye and without any effort on your part.

This kind of automated vigilance doesn't just save you from potential disasters; it also gives you peace of mind. While you're focusing on growing your business, AI is quietly standing guard, ensuring that malicious actors never get the chance to slip through the cracks.

Tackling Insider Threats with AI: Watching the Inside, Too

Not all threats come from shadowy figures on the internet. Sometimes, the danger is much closer to home. Insider threats—whether from a disgruntled employee or an innocent mistake—can be just as damaging as external attacks. And let's face it: in a small business, where teams often feel more like family, these risks can be particularly tough to spot.

This is where AI steps in as your internal watchdog. By monitoring patterns in employee behavior, AI can detect unusual activity that might otherwise fly under the radar. Imagine an employee suddenly accessing sensitive files they've never needed before or transferring data to external drives late at night. AI tools can flag these anomalies in real time, giving you a heads-up to investigate before the situation spirals out of control.

For small businesses, where trust is part of the culture, having this extra layer of vigilance is invaluable. AI isn't about mistrusting your team—it's about ensuring that mistakes or bad intentions don't jeopardize everything you've worked so hard to build. Think of it as a safety net, quietly working behind the scenes to keep your business secure, no matter where the threat comes from.

Cutting Through the Noise: How AI Tackles Alert Fatigue

If you've ever dealt with traditional security tools, you know the drill: your inbox floods with alerts, most of which turn out to be false alarms. After a while, it's easy to become desensitized, brushing off notifications without a second glance. The result? A real threat could slip through unnoticed amidst the chaos—a phenomenon aptly named "alert fatigue."

AI changes the game. Instead of inundating you with every potential issue, AI learns to separate the wheat from the chaff. By analyzing patterns and behaviors over time, it gets better at distinguishing genuine threats from harmless anomalies. The result? A dramatic reduction in false positives and a sharper focus on the alerts that actually matter.

For small business owners juggling countless priorities, this is a game-changer. You won't be bogged down with unnecessary noise, leaving you free to concentrate on running your business while staying confident that the AI has your back. It's like upgrading from a smoke detector that goes off every time you burn toast to one that only sounds the alarm when there's a real fire.

A Real-World Win: How AI Protected a Small Marketing Firm

Take the case of a boutique marketing firm that decided to level up its defenses by adopting an AI-powered cybersecurity tool. One evening, as the owner was winding down, the tool flagged a login attempt from a location halfway across the globe—a place the company had zero connections to. Recognizing the anomaly, the AI immediately blocked the attempt and sent a notification to the owner.

A closer look revealed the source of the breach: an employee had fallen victim to a phishing scam, and their credentials had been stolen.

Without the AI tool in place, the attacker could have gained access to sensitive client data, potentially causing financial havoc and reputational fallout. Instead, the quick response ensured the firm's systems stayed secure, and the owner could rest easy knowing the threat had been neutralized.

This isn't just a flashy success story—it's a testament to how AI is transforming cybersecurity for businesses of all sizes. It's not about luxury; it's about necessity. For small businesses navigating an increasingly treacherous digital landscape, AI isn't just keeping pace with threats—it's outsmarting them. The result? A more secure, more resilient business that's ready to take on whatever the future holds.

Improving Your Incident Response with AI

When a cyber incident strikes, time doesn't just tick—it sprints. How quickly you react can determine whether you're dealing with a manageable hiccup or a full-blown crisis. For small businesses, where resources and manpower are often stretched to the limit, speed isn't just an advantage—it's a lifeline.

Why Fast Response is Critical

Picture a cyberattack as a small fire. Caught early, it's a nuisance that can be extinguished with minimal fuss. But left unattended, it spreads rapidly, consuming everything in its path. In the digital world, this translates to stolen data, crippled systems, and reputational damage that could take years to rebuild.

For small businesses, the stakes are even higher. With fewer resources to absorb the impact of a breach, every second counts. That's where AI-powered tools shine. Unlike traditional systems that rely on humans to sift through logs and alerts, AI works at lightning speed, detecting threats in real time and, in many cases, taking immediate action to neutralize them.

How AI Speeds Up Incident Response

AI doesn't just identify threats—it acts on them. When an anomaly is detected, such as unauthorized access or suspicious file transfers, AI can block the activity, quarantine affected systems, and notify key personnel—all within seconds. This automated efficiency ensures that minor issues don't spiral into major disasters, giving small businesses a fighting chance to stay secure.

In today's fast-paced digital environment, response time is everything. AI-powered tools don't just buy you time—they give you the edge you need to protect your operations, your data, and your reputation. And for small businesses, that edge could mean the difference between thriving and surviving.

How AI-Driven Tools Can Automate Responses

Gone are the days when handling a cyber incident required waiting for IT staff to jump in, troubleshoot, and resolve the problem manually. AI-powered tools are redefining incident response by acting immediately—often before you even realize there's an issue. Here's how:

Blocking Access

Imagine an unauthorized login attempt from an unfamiliar location. Before the attacker can take another step, AI detects the anomaly and slams the virtual door shut, blocking access entirely. No waiting, no deliberation—just swift action to keep your systems safe.

Isolating Infected Devices

If malware finds its way onto a computer in your network, AI doesn't wait for a meeting or manual intervention. It isolates the infected device instantly, quarantining it to prevent the malware from spreading to other systems.

Alerting Key Personnel

While AI is busy containing the threat, it's also looping you and your team into the situation. Automated alerts provide real-time updates, ensuring you're informed and ready to take any additional steps required.

These tools excel in high-pressure situations where every second matters. By acting quickly and decisively, AI not only minimizes the potential damage but also removes the risk of human error—a common vulnerability when stress levels are high.

For small businesses, where IT resources are often limited, these AI-driven responses offer a vital safety net, delivering both speed and precision when it matters most.

Smarter, Not Just Faster: AI's Role in Incident Response

AI doesn't just sprint into action—it does so with a level of intelligence that transforms how small businesses handle cybersecurity threats. By automating critical incident response tasks, AI frees you from spending hours diagnosing problems and figuring out what to do next. Here's how it works:

Identifying Affected Systems

The moment a cyber incident occurs, AI dives into your network to pinpoint exactly which devices or accounts have been compromised. No more guessing games or hours of manual checks—AI shines a spotlight on the problem areas instantly.

Determining the Severity of an Attack

Not every threat is a four-alarm fire. Some are minor glitches, while others pose serious risks to your data and systems. AI evaluates the scope and severity of each attack, giving you a clear picture of whether you're dealing with a nuisance or a crisis.

Recommending Steps to Contain the Threat

AI doesn't just identify problems; it provides actionable solutions. Whether it's disabling a compromised account, patching a software vulnerability, or restoring data from a backup, AI delivers tailored recommendations to stop the threat in its tracks.

By autonomously handling these time-intensive tasks, AI empowers small business owners to stay focused on what they do best—running their businesses. Meanwhile, their AI-powered security systems quietly and efficiently neutralize threats behind the scenes.

For small businesses without the luxury of full-time IT staff, this smart, automated approach is a game-changer, offering the kind of protection that used to be reserved for enterprise-level operations.

AI-Powered SIEM: A Watchtower for Your Cybersecurity

AI is revolutionizing how businesses, large and small, monitor and respond to potential security threats. One of the standout tools in this transformation is **Security Information and Event Management (SIEM)** systems. These platforms act like the watchtower of your cybersecurity fortress, keeping an eye on everything that happens across your network and analyzing vast amounts of security data in real time.

How SIEM Works

SIEM tools gather logs and event data from across your systems—everything from login attempts and file transfers to software updates and access patterns. Using AI, these systems sift through the noise to identify patterns and detect anomalies that could indicate a threat. For example, if an unusual spike in failed login attempts occurs, a SIEM system can flag it as a potential brute force attack.

Why SIEM Matters for Small Businesses

Historically, SIEM systems were the domain of large corporations with big budgets and dedicated IT teams. But times have changed. Modern SIEM platforms are increasingly accessible, offering:

- **User-Friendly Interfaces**: You don't need a computer science degree to understand what's happening on your network.

- **Affordable Pricing**: Many solutions are now designed with small businesses in mind, offering scalable options that won't break the bank.

- **24/7 Monitoring**: Even when you're off the clock, your SIEM system is hard at work, ensuring no threat goes unnoticed.

Swift and Effective Responses

The real power of SIEM lies in its ability to not only detect threats but also respond to them quickly and effectively. AI-powered SIEM systems can trigger automated actions—such as isolating a compromised account or alerting you to unusual activity—within moments of detecting a potential issue.

For small businesses without a dedicated IT team, SIEM tools offer peace of mind, providing a level of monitoring and response that would otherwise be out of reach. With AI-powered SIEM in your corner, your cybersecurity defenses remain vigilant and robust, ensuring your business stays protected around the clock.

The Power of AI in Action: A Real-World Scenario

Picture this: A small but thriving design agency, built on the foundation of client trust and the careful handling of creative project files, faces a sudden and unexpected challenge. Late one evening, while the team is off the clock, their **AI-powered security tool** detects something unusual—a large file transfer is being initiated from a suspicious IP address.

The AI doesn't waste a second. Recognizing the anomaly, it swiftly isolates the affected device, effectively halting the unauthorized transfer and preventing any sensitive data from being stolen. Simultaneously, it alerts the agency owner, who logs in to find a detailed incident report waiting:

- The suspicious IP address has been traced to a phishing attack.

- The compromised employee account has been temporarily disabled to stop any further misuse.

- Recommendations for next steps, like resetting passwords and reviewing access logs, are laid out clearly.

The entire incident is resolved within minutes. No data is lost, and the agency's operations remain largely uninterrupted.

The AI Advantage

Without the AI-powered tool in place, this attack might have gone unnoticed for days—or even weeks—until significant damage had already been done. Sensitive client data could have been stolen, resulting in damaged reputations, broken trust, and potential legal ramifications.

Instead, the agency's proactive approach to cybersecurity, supported by AI, turned what could have been a devastating breach into a minor hiccup. The result? Continued client confidence, seamless operations, and the reassurance that their defenses are ready for whatever comes next.

AI isn't just a luxury—it's a lifeline for small businesses that want to stay secure in an increasingly digital world.

Why AI is a Game-Changer for Incident Response

AI doesn't just speed up incident response—it elevates it to an entirely new level of precision and effectiveness. For small businesses, this is a game-changer. It means that instead of playing catch-up with cyber threats, you're proactively staying one step ahead. With AI as part of your cybersecurity strategy, you're no longer relying on outdated tools or manual interventions. Instead, you're leveraging smart, efficient, and adaptive defenses that can handle threats swiftly and decisively.

Integrating AI into your cybersecurity approach isn't just about keeping up—it's about confidently leading the charge in an ever-evolving digital world. It's the assurance that your business is protected, your operations are secure, and your team can focus on growth rather than constantly worrying about what might go wrong. In the battle against cyber threats, AI ensures you're not just surviving—you're thriving.

The Future of AI in Cybersecurity

As we peer into the future, one thing is certain: artificial intelligence is poised to redefine the cybersecurity landscape. With its unparalleled ability to learn from patterns, adapt to new threats, and respond at lightning speed, AI is emerging as more than just a tool—it's a trusted ally. For small businesses, this evolution brings both challenges and opportunities. The road ahead promises advancements that will make defenses smarter, more accessible, and even more proactive, leveling the playing field in ways that once seemed impossible.

Let's dive into what lies ahead for AI in cybersecurity and explore how small businesses can prepare to embrace these cutting-edge solutions, ensuring they remain resilient in an ever-changing digital world. The future is bright—and armed with AI, it's one you can face with confidence.

How AI is Evolving in the Cybersecurity Field

AI technology is racing forward at a breathtaking pace, and with it, the landscape of cybersecurity is being reshaped. What once began as relatively simple tools for spotting anomalies has evolved into advanced systems capable of not only identifying and analyzing threats but also neutralizing them—all without human intervention.

Looking ahead, AI will shift from reacting to threats to predicting them with astonishing precision. Imagine a system so advanced it can anticipate a cyberattack before it even begins, giving businesses the power to act preemptively. For small businesses, this is a game-changer. As AI continues to mature, the gap between what's available to large corporations and smaller operations will shrink, granting access to defenses that are not only smarter and faster but also affordable and intuitive.

The future isn't just about keeping up—it's about staying ahead. For businesses of every size, AI offers a glimpse of a more secure tomorrow.

AI and Predictive Threat Intelligence

One of AI's most exciting talents lies in predictive threat intelligence. By sifting through vast amounts of historical data, spotting emerging patterns, and keeping tabs on evolving trends, AI can foresee potential cyber threats before they fully form. It's a bit like having a meteorologist who warns you of a storm days in advance—except this storm is made of malicious code, and the stakes are far higher.

Imagine a new type of ransomware making its rounds globally. AI systems, constantly tuned into the digital pulse, can detect its rise and send early warnings, prompting businesses to bolster their defenses against similar attacks. For small businesses, this isn't just helpful—it's revolutionary. Instead of scrambling to react after an incident, you'll be proactively shoring up your systems, minimizing risks, and avoiding costly downtime.

With AI, the game shifts from reacting to anticipating, giving you the power to stay one step ahead of cybercriminals.

AI and Machine Learning in Adaptive Security

Imagine a cybersecurity system that doesn't just defend but learns and grows with each passing day, adapting to stay ahead of increasingly cunning adversaries. This is the magic of adaptive security, made possible by machine learning.

Machine learning empowers AI to analyze vast troves of data, drawing lessons from past cyber incidents to refine its defenses. Picture a phishing attack that employs a new, creative twist. An adaptive AI system won't just block the immediate threat; it will analyze the technique, adjust its algorithms, and be ready to thwart similar attempts in the future. It's like hiring a security guard who gets sharper and savvier every time someone tries to outsmart them.

For small businesses, this self-improving capability is a game-changer. It's the assurance that your defenses don't just stand still—they evolve alongside the ever-changing tactics of cybercriminals, ensuring that your business stays protected no matter what's on the horizon.

The Role of AI in Combating AI-Driven Cyber Attacks

The cybersecurity battlefield is fast evolving into a high-stakes game of "AI vs. AI." Cybercriminals are now deploying AI to craft attacks that are more sophisticated, adaptive, and insidiously clever. We're talking about malware that learns to dodge detection or phishing scams so eerily human-like, they could fool even the sharpest eye.

In this escalating arms race, AI-powered defenses are no longer optional—they're essential. The tools of tomorrow will need to outsmart malicious AI, spotting subtle patterns and responding with precision in real time. This shift underscores the dual-edged nature of AI: it's a powerful ally for defenders but an equally formidable weapon for attackers.

Small businesses might feel daunted by this development, but here's the encouraging truth: you don't have to go it alone. With the right AI tools in your corner, even a lean team can hold its own, standing toe-to-toe with cybercriminals and keeping systems secure against increasingly cunning threats. In this AI-driven era, it's not about the size of your team—it's about the strength of your defenses.

Future AI-Powered Cybersecurity Tools for Small Businesses

The future of AI in cybersecurity holds exciting possibilities, particularly for small businesses. Emerging tools are becoming more accessible and affordable, promising features like:

- **Advanced Behavioral Analytics**: AI systems that can understand individual user behavior, spotting irregularities before they become a problem.

- **Fully Automated Incident Response**: Tools capable of handling the entire lifecycle of a cyber incident, from detection to resolution, without requiring human intervention.

- **Personalized Security Recommendations**: AI platforms that analyze a business's unique needs and provide tailored strategies for maximum protection.

As these tools become more mainstream, small businesses will have the opportunity to deploy enterprise-grade defenses on a fraction of the budget.

Preparing for an AI-Driven Cybersecurity Future

Staying ahead in the AI-driven cybersecurity landscape isn't just about defense—it's about preparation and adaptability. For small businesses, keeping an eye on emerging AI technologies and trends is vital. These advancements aren't just shiny tech upgrades; they're tools that can significantly bolster your security posture.

Investing time now to understand AI-powered solutions pays dividends in the long run. By staying informed, you ensure your business is ready to counter ever-evolving threats while remaining agile enough to seize the opportunities that advanced technology brings. Think of it as future-proofing—not just for your cybersecurity, but for the longevity and resilience of your business.

Looking Ahead

The future of cybersecurity and AI are intertwined in a way that's both exciting and empowering. For small businesses, this isn't just a challenge to navigate—it's a chance to level the playing field. AI is no longer a tool reserved for tech giants; it's an opportunity for businesses of all sizes to build smarter, more adaptive defenses.

By embracing these advancements, you're not just safeguarding your business against today's threats—you're positioning it to thrive in an increasingly digital world. Adopting AI-powered security isn't just about protection; it's about future-proofing your business for whatever comes next.

Implementing AI-Powered Defenses in Your Business

AI-powered cybersecurity tools might sound like something out of a sci-fi movie, but the reality is far more approachable—and affordable—than you might think. Implementing them in your business doesn't require an army of IT specialists or a Hollywood budget. The key lies in understanding your unique needs and taking a gradual, step-by-step approach to integration. Let's explore how to get started with confidence.

1. **Assess Your Current Defenses:** Begin by evaluating your existing cybersecurity setup. What tools are you already using? Where are your vulnerabilities? Are you struggling with phishing attacks, insider threats, or managing overwhelming alerts? Understanding your pain points will help you prioritize which AI-powered tools to explore first.

2. **Start Small and Build Gradually:** There's no need to overhaul your entire system overnight. Start by integrating AI into one area, like threat detection or email security. Tools such as AI-driven spam filters or anomaly detection software are often affordable and easy to implement, providing immediate value without a steep learning curve.

3. **Focus on Compatibility:** Choose AI tools that align with your current systems. Many modern solutions are designed

to integrate seamlessly with popular platforms, whether it's your email provider, cloud storage, or existing antivirus software. This ensures a smoother adoption process without unnecessary disruption.

4. **Leverage Free Trials and Demos:** Many AI-powered tools offer free trials or demos, allowing you to test their features before committing. Use these opportunities to explore how the tool fits your needs, and don't hesitate to ask providers for guidance on setup and customization.

5. **Train Your Team:** Even the best tools are only as effective as the people using them. Provide your team with simple, hands-on training to ensure they understand how to interact with AI systems and recognize when the tool is alerting them to a threat.

By taking these manageable steps, you'll demystify AI and make it a natural part of your business's cybersecurity strategy. AI doesn't have to be overwhelming—it's a tool, not a magic wand, and when implemented thoughtfully, it can elevate your defenses while fitting seamlessly into your existing workflows.

Assessing Your Business Needs

Not all businesses face the same cybersecurity challenges, which is why your choice of tools should align with your specific risks and priorities. Before diving into AI-powered solutions, it's essential to evaluate your current cybersecurity posture. Ask yourself the following questions:

1. **What data or systems are most critical to your operations?** Identify the assets that, if compromised, would have

the most significant impact on your business. For example, customer payment information, proprietary designs, or confidential client records might top your list.

2. **What types of threats are you most likely to encounter?** Consider the threats your business is most exposed to. Phishing attacks, ransomware, insider threats, or even supply chain vulnerabilities may be more relevant depending on your industry and operations.

3. **Do you have in-house IT expertise, or do you rely on external support?** This will help determine whether you need user-friendly tools designed for non-specialists or more advanced solutions that might require some technical know-how.

Honestly, you would do yourself a favor if you asked the professionals to do an assessment for you. Even if it cost a few bucks (probably more like $1000) plus, you'll end up with a better foundation and a much better understanding of where you currently lie on the vulnerability spectrum.

Turning Insights into Action

By answering these questions, you can identify gaps in your defenses and focus on the tools that will deliver the most value. For instance:

- **A small e-commerce business** might prioritize tools to detect payment fraud or enhance secure online transactions.

- **A creative agency** may benefit more from AI-powered file protection and malware scanning to safeguard client work.

- **A healthcare provider** could focus on tools to ensure compliance with HIPAA and protect sensitive patient data.

Cybersecurity isn't a one-size-fits-all solution. By tailoring your approach to your specific needs, you can make smarter investments in tools that provide the most meaningful protection for your business.

Choosing the Right AI-Powered Security Tools

Once you've identified your cybersecurity needs, the next step is finding AI-powered tools that align with your budget and priorities. The good news? These tools come in a variety of forms, designed to address specific challenges and scale to meet the demands of small businesses.

1. Automated Threat Detection

These tools act as tireless watchdogs, scanning your systems in real-time for suspicious activity. Whether it's an unusual login attempt or a flood of unexpected data transfers, automated threat detection identifies and neutralizes issues before they escalate.

2. SIEM Solutions

Security Information and Event Management (SIEM) systems are the Swiss Army knife of cybersecurity tools. They collect, consolidate, and analyze security data from across your network, providing actionable insights to help you detect and respond to threats faster. Many modern SIEM tools include AI features, making them both efficient and user-friendly for small businesses.

3. AI-Powered Malware Scanners

Malware is one of the most common threats faced by businesses of all sizes. AI-powered malware scanners automatically detect and block malicious software—like ransomware or spyware—before it can compromise your systems. These tools adapt to new threats, offering a proactive layer of protection.

Balancing Features and Affordability

For small businesses, affordability is a critical factor. Here's how to make the most of your investment:

- Look for tools with **flexible pricing plans** that let you pay for only the features you need.

- Take advantage of **free trials** or freemium versions to test solutions before committing.

- Focus on tools designed with small businesses in mind—they're often simpler to use and prioritize cost-efficiency.

By selecting AI-powered tools tailored to your specific needs and budget, you can achieve robust cybersecurity without overstretching your resources. These tools provide a strong starting point, allowing you to build a scalable, future-ready defense strategy.

Integrating AI Tools with Existing Security Measures

AI is a game-changer in cybersecurity, but it's most effective when used as part of a broader, layered defense strategy. Rather than replacing your existing security measures, AI tools amplify and enhance them, creating a more robust shield against cyber threats. To get the best results, integrate AI with these foundational components:

1. Firewalls: The Frontline Defender

Firewalls act as your first line of defense, controlling traffic in and out of your network. They block unauthorized access and filter potentially harmful data. AI tools work alongside firewalls by analyzing traffic patterns and identifying anomalies that might signal an attack.

2. Antivirus Software: A Trusted Ally

Traditional antivirus programs focus on known threats, while AI tools excel at identifying emerging risks in real time. Together, they form a powerful duo—antivirus software handles the basics, while AI steps in to detect and respond to sophisticated attacks like zero-day exploits or novel malware.

3. Employee Training: The Human Factor

Even the most advanced AI systems can't prevent human mistakes. Phishing emails, weak passwords, and accidental clicks remain some of the biggest vulnerabilities in any organization. Regularly train your team to recognize scams, avoid risky behaviors, and follow cybersecurity best practices. An informed workforce significantly reduces the chances of an incident.

Creating a Seamless Security Ecosystem

Integration is the key to making all these components work together effectively. By aligning your AI tools with firewalls, antivirus software, and employee training programs, you ensure that your defenses are not only comprehensive but also cohesive. AI doesn't just fill gaps—it strengthens your overall security posture, making your business more resilient to threats.

The result? A balanced, layered cybersecurity strategy where each element supports the other, leaving attackers with fewer opportunities to exploit.

Getting Started with AI-Powered Solutions

Incorporating AI into your business's cybersecurity doesn't have to be an overwhelming, all-at-once transformation. A gradual, step-by-step approach lets you reap immediate benefits without stretching your budget or resources too thin. Here's how to take those first steps with confidence:

1. Consult an IT Professional

Feeling uncertain about where to start? You're not alone. Cybersecurity is a complex field, and an IT professional can act as your guide. They can assess your current defenses, identify your business's unique vulnerabilities, and recommend AI tools that align with your needs and budget. This expert advice helps you avoid wasting time or money on solutions that don't fit.

2. Implement in Stages

There's no need to overhaul your entire system overnight. Start small by integrating a single AI-powered tool into your existing defenses. For instance, an AI-powered email security system can immediately reduce your exposure to phishing attacks—a common threat for small businesses. Once you're comfortable with that tool, consider expanding to others, like real-time threat detection software or AI-enhanced antivirus solutions.

3. Monitor and Adjust

As you add each tool, track its impact and performance. Are incidents decreasing? Is your team finding the tools user-friendly? Use this feedback to fine-tune your strategy before moving on to the next stage.

Why a Gradual Approach Works

By starting small, you gain valuable hands-on experience with AI-powered solutions without overwhelming your team or budget. This phased method allows you to build confidence in the technology, understand how it integrates with your existing systems, and make adjustments along the way.

In the end, implementing AI is less about a dramatic overhaul and more about thoughtful, incremental progress. Each step strengthens your defenses, proving that even small investments in AI can yield significant rewards.

Fine-Tuning Your AI-Powered Defenses

Implementing AI-powered cybersecurity tools is a strong first step, but keeping them effective requires regular attention. Think of these tools not as "set it and forget it" systems, but as dynamic components that need ongoing care to adapt to the ever-changing cyber threat landscape. Here's how to make sure your AI defenses are always working at their best:

1. Track Performance Regularly

Most AI tools come equipped with robust reporting features. Use these to monitor how well they're detecting and mitigating threats. Are the tools flagging genuine risks? Are they reducing the number of incidents? Reviewing these metrics not only confirms their effectiveness but also highlights areas for improvement.

2. Adapt to Changes

As your business evolves—whether you're adding new employees, launching a new product, or expanding into different markets—your cybersecurity needs will shift. Similarly, cyber threats are always evolving. Regularly review your AI configurations and update them to address new vulnerabilities and challenges.

3. Gather Team Feedback

Your employees are on the front lines of using these tools. Encourage them to share their experiences—whether it's a seamless feature they appreciate or a hiccup that causes frustration. Their feedback can provide invaluable insights into what's working and what might need tweaking.

Why It Matters

AI-powered defenses are designed to adapt and learn, but they're only as effective as the environment they're working in. By regularly reviewing and fine-tuning these tools, you ensure they remain aligned with your business's needs and resilient against emerging threats.

Think of this process as ongoing maintenance—like tuning up a car to keep it running smoothly. With regular monitoring and adjustments, your AI defenses will continue to provide robust protection, giving you peace of mind in an unpredictable digital world.

A Smarter, Safer FutureBy thoughtfully implementing AI-powered cybersecurity tools, you're not only safeguarding your business today but also building a foundation to tackle the challenges of tomorrow. The digital landscape is constantly evolving, and AI offers the agility and intelligence needed to keep pace with emerging threats.

The process begins with understanding your unique vulnerabilities and priorities. By identifying the areas where your business is most at risk, you can select tools that provide the greatest impact. From there, integration is key—combining AI with your existing defenses to create a cohesive, layered strategy.

But implementation is just the start. Regular monitoring, adaptation, and an openness to evolving technologies ensure that your defenses remain effective as new challenges arise. AI is not a static solution; it's a dynamic ally that learns, grows, and adjusts alongside your business.

The result? A business that's not only protected but also empowered to thrive in an unpredictable digital world. With AI on your side, you're transforming cybersecurity from a reactive necessity into a proactive advantage. It's an investment in resilience, trust, and peace of mind—today and into the future.

Setting Realistic Expectations for AI-Powered Defenses

AI is a powerful ally, but it's not a magic wand. While it excels at detecting and responding to threats at speeds no human can match, it's most effective as part of a broader, well-rounded cybersecurity strategy.

- **AI Can't Replace Vigilance**: Employee training remains crucial. Your team still needs to recognize phishing attempts, maintain strong passwords, and report suspicious activity.

- **AI Needs a Backup Team**: Traditional defenses—like firewalls, antivirus software, and system updates—are essential companions to AI tools. Think of AI as an enhancer, not a replacement.

- **AI Requires Fine-Tuning**: Straight out of the box, AI tools might not operate at their peak. Regular monitoring and adjustments will ensure they adapt to your business's specific needs and evolving threats.

Understanding what AI can—and cannot—do helps you strike the right balance, allowing you to leverage its strengths without over-relying on it.

Staying Informed: Keeping Up with AI in Cybersecurity

AI technology is advancing at lightning speed, and staying updated is crucial for maintaining effective defenses. Here's how to stay ahead:

- **Follow Trusted Sources**: Subscribe to newsletters, blogs, or podcasts from reputable organizations like the Cybersecurity and Infrastructure Security Agency (CISA), *Wired*, or industry leaders like Gartner.

- **Engage with Online Communities**: Join forums or social media groups where small business owners and cybersecurity professionals exchange insights on tools and trends.

- **Attend Webinars and Events**: Take advantage of free webinars hosted by cybersecurity companies to learn about emerging technologies and best practices.

- **Monitor Your Tools**: Keep an eye on updates and new features from your existing AI tools. Vendors frequently release improvements to counter the latest threats.

Staying informed ensures you're not just reacting to changes in cybersecurity but proactively adapting your defenses to meet new challenges.

Moving Forward with Confidence

Integrating AI into your cybersecurity strategy is more than just a smart move—it's a commitment to safeguarding your business's future. By aligning AI tools with your overall strategy, maintaining realistic expectations, and staying informed about advancements, you're positioning your business to thrive in an increasingly digital world.

With AI as your ally, you can face evolving cyber threats with confidence and focus on what truly matters: growing your business and delivering for your customers. The digital frontier doesn't have to be daunting when you've got the right tools—and the right mindset—on your side.

Wrapping Up: The Game-Changing Role of AI in Small Business Cybersecurity

As we near the end of this chapter, it's crystal clear that artificial intelligence isn't just a buzzword—it's a transformative tool reshaping the cybersecurity landscape. For small businesses, AI-powered solutions present a remarkable opportunity to level the playing field, equipping you with defenses that once seemed exclusive to tech giants and multinational corporations.

AI: The Key to Small Business Resilience

In the past, robust cybersecurity measures often felt out of reach for small businesses, hindered by budget constraints and limited IT expertise. But AI has rewritten that narrative, offering tools that are not only powerful but also accessible and scalable. Whether it's real-time threat detection, anomaly analysis, or automated incident response, AI brings unprecedented capability to businesses of every size.

For small businesses, the power of AI is about more than just protection—it's about empowerment. It enables you to preemptively tackle threats like ransomware and phishing, preventing damage before it happens. AI doesn't just help you keep up with cybercriminals; it allows you to get ahead of them, creating a secure foundation for your operations and your reputation.

Adaptation and Proactivity: Staying Ahead of the Curve

Cybersecurity is a constantly shifting game, with cybercriminals refining their tactics as technology evolves. AI's strength lies in its ability to adapt and learn over time, growing stronger with each interaction. However, leveraging AI effectively requires an active, informed approach.

Stay engaged. Stay proactive. Explore new tools, attend webinars, and regularly evaluate your cybersecurity strategy as your business grows. Cybersecurity isn't static, and neither can your defenses be. By treating AI as an evolving partner rather than a static solution, you'll ensure your business remains secure against emerging threats.

Navigating the Complexity of AI Implementation

While AI offers small businesses immense potential, it's important to approach its implementation with realism. Tools like Security Information and Event Management (SIEM) systems and advanced behavioral analytics platforms are powerful, but their full potential often depends on precise configuration and ongoing management—tasks that may exceed the capabilities of smaller in-house teams.

This is where expert assistance shines. Cybersecurity consultants and Managed Service Providers (MSPs) specialize in deploying these tools effectively. They can tailor AI-driven solutions to meet your business's specific needs, minimizing false positives, ensuring seamless integration, and optimizing performance.

Investing in professional support isn't a sign of weakness; it's a strategic move that ensures you're building your defenses on a solid foundation. Think of it as enlisting a guide to navigate the complex terrain of AI-powered cybersecurity. The peace of mind it brings far outweighs the upfront cost.

Preparing for the Final Chapter

As we transition to the final chapter, we'll tie together the strategies, tools, and insights from every step of this journey. You'll leave with a comprehensive roadmap to implement a cybersecurity plan that's both effective and sustainable. From the foundational steps to advanced AI-driven solutions, the next chapter will ensure you're equipped to protect your business now and into the future.

AI isn't just a tool—it's an opportunity to redefine how small businesses approach security. By embracing its potential and staying committed to adaptation, you're setting your business up for a future where it's not just protected, but thriving. The final chapter will show you how to make that vision a reality. Stay tuned!

PREPARING FOR THE FUTURE

A CYBERSECURITY PLAN FOR SMALL BUSINESSES

Wrapping Up the Journey: Building a Resilient Future in Cybersecurity

C ybersecurity is not a finish line you cross; it's a journey that evolves with every new threat, technology, and milestone your business encounters. As we step into this final chapter, let's focus on weaving together everything you've learned to prepare for the road ahead. This isn't just about surviving—it's about thriving in a digital world that grows more complex each day.

Reflecting on How Far You've Come

Take a moment to look back at the progress you've made. What once may have seemed like a daunting, jargon-filled topic now feels more manageable, and that's no small accomplishment. You've navigated through the maze of cybersecurity threats, from phishing scams to ransomware, and explored cutting-edge tools like AI that were once reserved for tech giants. Along the way, you've discovered practical strategies tailored to small businesses—tools, tips, and insights that empower you to take control of your digital defenses.

When you first started this journey, cybersecurity might have felt like a job for someone else—a mysterious realm best left to IT experts. But now, you're equipped with a deeper understanding of the risks, the resources, and the ways to build a defense system that fits your business. And while you're likely more confident than before, the key takeaway is this: cybersecurity isn't something you do once and forget. It's a commitment.

Embracing a Long-Term Mindset

Think of cybersecurity like fitness. You don't hit the gym for a month and expect to stay in shape forever. Similarly, your digital defenses need ongoing care and attention to remain effective. Cybercriminals are constantly evolving their tactics, using emerging technologies—sometimes even AI—to exploit vulnerabilities. Meanwhile, as your business grows, your security requirements will expand alongside it. What worked for your business yesterday may not be enough tomorrow.

Here's what embracing a long-term cybersecurity mindset means:

- **Regular Check-Ups**: Just like annual health check-ups, your systems, policies, and tools need periodic reviews. Are your defenses still strong? Are there new tools or updates you should consider?

- **Team Awareness**: A culture of cybersecurity begins with your people. Employees who are trained to recognize and respond to threats are your greatest asset.

- **Adaptability**: Stay informed about new technologies and threats, and be prepared to adapt your strategy. What seems like overkill today may be standard practice tomorrow.

This mindset ensures that cybersecurity becomes a natural part of your daily operations—not a separate, overwhelming task to tackle when disaster strikes.

The Journey, Not the Destination

It's tempting to view cybersecurity as a checklist of tasks you can complete and move on from, but that's not how this works. Cybersecurity is dynamic—it's less about building an impenetrable fortress and more about staying vigilant, proactive, and adaptable. The goal isn't perfection; it's resilience. It's about bouncing back quickly from potential threats and continuously strengthening your defenses.

Small businesses don't need the resources of multinational corporations to stay secure. What they need is a plan, the right tools, and the commitment to keep learning and growing. The beauty of the journey is that every small step you take—training employees, updating software, or integrating AI—makes a big difference over time.

A Comprehensive Plan for the Future

This chapter will guide you in creating a holistic cybersecurity plan that draws on all the lessons you've learned in this book. Together, we'll map out:

- How to assess and prioritize your risks

- The tools and strategies to protect your business effectively

- A schedule for maintaining and improving your defenses

- Ways to stay informed about emerging threats and technologies

This plan isn't just a checklist—it's a living, adaptable strategy that grows with your business. By the end of this chapter, you'll have a roadmap for navigating the evolving cybersecurity landscape with clarity, confidence, and control.

Looking Forward

Cybersecurity may feel like a challenge, but it's also an opportunity—a chance to build a foundation of trust, safeguard your hard work, and future-proof your business. The road ahead will have its twists and turns, but with the right tools and mindset, you're more than ready to tackle them. Let's take this final step together and create a plan that ensures your business isn't just surviving in a digital world—it's thriving.

Revisiting the Threat Landscape: A Recap of Major Cybersecurity Risks

As we approach the end of this book, it's time to reflect on the key threats small businesses face in today's digital world. These challenges, while daunting, are manageable with the right preparation, tools, and mindset. Here's a summary of the major risks we've covered and why addressing them proactively is essential.

Social Engineering: The Human Factor

Perhaps the most insidious of cyber threats, social engineering exploits human nature—our trust, curiosity, and desire to help. Phishing tops the list, tricking individuals into clicking malicious links or divulging sensitive information. Pretexting takes this a step further, with attackers posing as trusted figures like vendors or coworkers to gain access.What makes social engineering so dangerous is its subtlety. Even diligent employees can fall for well-crafted schemes. That's why combating it requires a twofold approach: comprehensive employee training to recognize red flags and robust technological safeguards to catch what humans might miss.

Malware and Ransomware: A Double Threat

For small businesses, malware and ransomware can be catastrophic. Malware infiltrates systems to steal data or disrupt operations, while ransomware locks you out of your own systems until a ransom is paid. Both are costly—not just in financial terms but in lost customer trust and reputational damage.Small businesses are often seen as low-hanging fruit by attackers, making defenses like regular backups, strong endpoint security, and prompt software updates critical. These simple steps can make the difference between a minor inconvenience and a major crisis.

Insider Threats: The Danger Within

While we often picture cyberattacks coming from shadowy figures in far-off locations, insider threats are a significant concern. These can be intentional—like a disgruntled employee misusing access—or accidental, such as an employee falling for a phishing email and inadvertently compromising your systems.The solution? Foster a culture of cybersecurity awareness, monitor internal activities for unusual behavior, and implement the principle of least privilege, ensuring employees have access only to what they need to do their jobs.

Weak Passwords and the Lack of MFA: Simple Problems, Big Risks

It's staggering how many breaches boil down to something as basic as weak passwords. A single vulnerable password can serve as an open door for attackers. Multi-factor authentication (MFA) is an easy-to-implement safeguard, requiring an extra verification step that stops unauthorized access—even if passwords are compromised. Establishing strong password policies and mandating MFA are some of the simplest, most cost-effective ways to boost your defenses.

AI-Powered Attacks: The New Frontier

As AI becomes an essential tool for defense, cybercriminals are using it to supercharge their attacks. AI enables them to craft highly convincing phishing emails, create adaptive malware, and automate attacks on an unprecedented scale.For small businesses, this presents a new layer of complexity. But it also underscores the importance of embracing

AI as part of your defensive strategy. AI-powered tools can detect patterns, analyze threats in real time, and respond faster than human teams alone, leveling the playing field against advanced attackers.

The Evolving Threat Landscape: Why These Risks Are Here to Stay

The methods may change, but the motivations—financial gain, disruption, and exploitation—remain constant. Cybercriminals are relentless in their innovation, finding new ways to bypass defenses and exploit weaknesses. For small businesses, the takeaway is clear: complacency is not an option.

While you can't predict every new tactic, you can prepare. Staying informed about emerging threats, adopting layered defenses, and leveraging technologies like AI are not just best practices—they're essential practices.

Building a Resilient Defense

Cybersecurity is not a finish line; it's an ongoing journey. Each of the threats we've discussed is a reminder that protecting your business requires commitment, vigilance, and adaptability. The good news? You now have the tools, knowledge, and strategies to navigate this landscape with confidence.

By focusing on proactive measures and maintaining a culture of security awareness, small businesses can transform these challenges into opportunities to build trust, strengthen operations, and safeguard their hard-earned success. Cyber threats are here to stay, but with the right approach, so is your resilience.

Your Cybersecurity Action Plan

Building a solid cybersecurity framework doesn't have to be over-whelming. By following a step-by-step action plan, small businesses can create a manageable, effective strategy to protect their operations. This section provides a detailed roadmap to help you assess your current cybersecurity posture, implement key measures, and maintain ongoing vigilance.

Step 1: Assess Your Business's Current Cybersecurity Posture

Before taking action, you need to understand where your business stands. Conducting a simple self-assessment can highlight both your strengths and areas for improvement. Here's how:

- **Check Existing Security Tools**: Inventory what you already have in place, such as firewalls, antivirus software, and access controls. Are these tools updated and functioning effectively?

- **Review Employee Training**: Assess your team's awareness of cybersecurity risks. Do they know how to recognize phishing emails, use strong passwords, or report suspicious activity?

- **Evaluate Compliance**: Ensure your business meets any industry-specific cybersecurity regulations or standards. For example, businesses handling customer payment data must comply with PCI-DSS.

This assessment serves as the foundation for prioritizing your next steps, so be honest about any gaps or weaknesses you uncover.

Step 2: Prioritize Key Cybersecurity Measures

Once you've identified your vulnerabilities, focus on implementing foundational measures that provide the biggest security benefits. These steps are essential for any business:

- **Strong Password Policies and MFA**: Require employees to use complex passwords and enable multi-factor authentication on all critical accounts. MFA significantly reduces the risk of unauthorized access.

- **Regular Software Updates**: Outdated software is a common entry point for cyberattacks. Schedule regular updates for operating systems, applications, and firmware.

- **Employee Training**: Empower your team to be your first line of defense by teaching them to recognize phishing attempts, avoid suspicious links, and report security concerns.

- **Data Backups**: Establish automatic backups for critical data and store copies offsite or in the cloud. This ensures you can recover quickly from ransomware or other data loss incidents.

Prioritizing these measures creates a strong cybersecurity foundation that protects your business from the most common threats.

Step 3: Create a Basic Incident Response Plan

Preparation is key to minimizing the impact of a cyber incident. A straightforward incident response plan ensures your business can respond quickly and effectively. Here's what to include:

- **Incident Management Roles**: Assign responsibility for managing cyber incidents to specific team members or external partners. Make sure they understand their roles and are trained for emergencies.

- **Isolation and Containment Steps**: Define the steps to isolate affected systems, such as disconnecting infected devices from the network or temporarily disabling compromised accounts.

- **Communication Channels**: Establish clear channels for internal communication during an incident. Employees should know who to contact and how to report issues.

- **Reporting Process**: Decide how incidents will be documented, including what details to include in reports and how they'll be used to improve defenses.

A simple, well-communicated plan can significantly reduce downtime and mitigate damage in the event of a breach.

Step 4: Implement Ongoing Cyber Hygiene Practices

Cybersecurity isn't a one-time task—it's a continuous effort. Encourage your team to adopt everyday habits that protect your business, such as:

- **Performing Regular Backups**: Ensure backups are automated and tested for reliability.

- **Keeping Systems Updated**: Schedule updates to address vulnerabilities promptly.

- **Practicing Good Digital Habits**: Avoid public Wi-Fi for sensitive work, use secure file-sharing methods, and maintain clean device management practices.

Building a culture of cybersecurity within your business ensures that best practices become second nature for everyone involved.

Step 5: Establish a Regular Review Process

Cyber threats evolve, and so should your defenses. Schedule regular reviews to evaluate the effectiveness of your cybersecurity measures:

- **Quarterly or Biannual Assessments**: Review your security tools, training programs, and incident response plan. Update them based on lessons learned and emerging threats.

- **Performance Metrics**: Track metrics such as the number of phishing attempts blocked, backup success rates, or response times during simulated incidents.

- **Adjustments**: Use the insights from reviews to refine your strategy and prioritize new investments in cybersecurity.

Regular check-ins help ensure your defenses remain strong and adaptable over time.

Sample Cybersecurity Action Plan for Small Businesses

Here's an example of how a small business might structure its cybersecurity action plan:

- **Monthly Tasks**:

  - Install and verify software updates.

  - Perform system and data backups.

  - Review security alerts and logs for unusual activity.

- **Quarterly Tasks**:

  - Conduct employee training sessions on phishing awareness and password policies.

- ○ Test incident response procedures with simulated scenarios.

- ○ Evaluate the performance of AI-powered security tools and make adjustments as needed.

- **Annual Tasks**:

  - ○ Perform a comprehensive security audit with the help of a professional.

  - ○ Review and update your incident response plan.

  - ○ Assess compliance with any industry regulations and address gaps.

This template can be tailored to suit your specific needs and resources, providing a practical framework to maintain and improve your cybersecurity defenses.

With a clear action plan in place, your business is well-equipped to navigate the ever-changing cybersecurity landscape. By focusing on foundational measures, preparing for incidents, and committing to ongoing vigilance, you're creating a safer, more resilient future for your operations

Innovation and Adaptability as the Key to Success

In the dynamic world of cybersecurity, standing still is not an option. Cyber threats evolve rapidly, and the technologies that protect against them must adapt just as quickly. For small businesses, the key to staying secure lies in embracing innovation and maintaining flexibility.

Let's explore why adaptability matters and how businesses can foster it, even on a limited budget.

Why Cybersecurity Requires Adaptability

Cyber threats are not static; they grow more sophisticated every day. What worked last year—or even last month—may not be sufficient to counter new tactics employed by cybercriminals. Small businesses face unique challenges in this landscape because they often have fewer resources to dedicate to cybersecurity.

Adaptability means being willing to evaluate your defenses regularly and adjust them to meet emerging challenges. It involves recognizing that no system is foolproof and staying open to new solutions that strengthen your security. Businesses that remain flexible are better positioned to anticipate threats, respond quickly, and minimize damage when incidents occur.

Staying Updated on Emerging Threats and Technologies

Staying informed is one of the most effective ways to stay ahead of cyber threats. The cybersecurity landscape is constantly changing, but small businesses can use these strategies to keep up:

- **Follow Trusted Cybersecurity Blogs**: Websites like Krebs on Security, Dark Reading, and the Cybersecurity and Infrastructure Security Agency (CISA) blog provide timely updates on threats and solutions.

- **Subscribe to Newsletters**: Many cybersecurity organizations offer newsletters with digestible insights and advice tailored to small businesses.

- **Attend Webinars and Conferences**: Webinars hosted by industry experts or security vendors can help you understand emerging technologies and trends. Many are free and designed to educate businesses of all sizes.

- **Use Social Media**: Platforms like LinkedIn and Twitter are excellent for following cybersecurity professionals who share insights and updates in real time.

By staying informed, you'll be better equipped to recognize new threats and make proactive decisions.

The Role of Continuous Learning in Cybersecurity

Cybersecurity isn't a one-and-done effort; it requires a commitment to ongoing education. Continuous learning ensures that you and your team remain prepared to face new risks. Here's how to make it a priority:

- **Host Regular Training Sessions**: Keep employees updated on topics like phishing awareness, secure password practices, and safe browsing habits.

- **Encourage Self-Education**: Provide access to online courses or resources for employees who want to deepen their understanding of cybersecurity.

- **Create a Learning Culture**: Foster an environment where employees feel encouraged to share insights and discuss potential threats they encounter.

A well-informed team acts as your first line of defense, significantly reducing the likelihood of human errors leading to breaches.

Investing in Innovation Within Budget Constraints

Innovation doesn't have to mean breaking the bank. Small businesses can implement creative and cost-effective strategies to enhance their security. Here's how:

- **Adopt Gradually**: Start with one AI-powered tool or security measure that addresses your most pressing vulnerabilities, such as an AI-driven phishing detection system or a password manager with MFA support. Expand as your budget allows.

- **Utilize Free Resources**: Many organizations, including CISA and NIST, provide free cybersecurity guides and tools tailored for small businesses.

- **Explore Affordable Tools**: Some vendors offer scalable solutions or entry-level plans that fit small business budgets without compromising on quality.

- **Optimize Existing Resources**: Maximize the effectiveness of the tools you already have by ensuring they're properly configured and maintained. For example, a properly updated firewall or antivirus solution can go a long way in reducing risks.

Innovation is as much about mindset as it is about technology. Even small changes, implemented thoughtfully, can have a significant impact.

Collaborating with Cybersecurity Professionals and Peers

You don't have to navigate the complexities of cybersecurity alone. Building relationships with experts and peers can provide valuable insights and resources. Here's how collaboration can help:

- **Partner with Professionals**: Cybersecurity consultants or managed service providers (MSPs) can help you assess your risks, implement tools, and maintain your defenses effectively. Even periodic consultations can make a big difference.

- **Join Peer Networks**: Online forums, LinkedIn groups, and local business associations often host discussions about shared challenges and solutions. These spaces allow you to exchange practical advice with other small business owners.

- **Participate in Workshops and Community Events**: Many organizations offer free or low-cost cybersecurity workshops where you can learn and connect with others facing similar challenges.

- **Share Resources**: Collaborating with peers to share knowledge about affordable tools, training resources, or best practices can stretch your budget further.

Building a network of trusted advisors and colleagues can provide you with the support and guidance needed to stay resilient in an ever-changing cybersecurity landscape.

The Bottom Line: Adaptability and Innovation Are Essential

In cybersecurity, the only constant is change. By staying adaptable, prioritizing continuous learning, and embracing innovation, small businesses can turn what might seem like a daunting challenge into a manageable—and even empowering—opportunity. Collaboration and creativity further enhance these efforts, ensuring that your business remains not only protected but also positioned for growth in an increasingly digital world.

A Final Call to Action

As we wrap up this journey through cybersecurity, one truth stands out: protecting your business is not just about tools or policies—it's about people. Cybersecurity is a shared responsibility, a collective effort where every small action contributes to a much larger defense. Whether you're the CEO, an employee, or an IT professional, your role matters in securing not only your business but also the broader digital ecosystem.

Cybersecurity as a Shared Responsibility

Imagine cybersecurity as a chain. Each person in your organization is a link, and the strength of the chain depends on everyone doing their part. From leadership setting the tone with strong policies to employees practicing good habits like recognizing phishing emails, everyone contributes to the safety of the business.

This shared responsibility doesn't require perfection—it requires participation. A team that understands the importance of vigilance, communicates openly about threats, and works together to follow best practices creates an environment where cyberattacks find fewer opportunities to succeed.

Empowering Small Businesses to Take Action

Small businesses are often seen as easy targets by cybercriminals, but they are far from powerless. With the right mindset, tools, and practices, small businesses can stand firm against even sophisticated threats.

You've learned that adopting AI-powered tools, setting strong passwords, enabling multi-factor authentication, and training your employees are all within your reach. These actions don't require an army of IT experts or an unlimited budget—they require commitment. By taking these steps, you're proving that small businesses can not only survive in the digital world but thrive with confidence.

Embracing the Role of Every Business in a Secure Digital World

Every business, no matter its size, has a role to play in creating a safer digital environment. Think of cybersecurity as a collective effort. When your business secures its systems, you're not just protecting your operations—you're protecting your clients, partners, and the broader community.

Imagine a world where every small business takes cybersecurity seriously. The ripple effect would create a digital landscape that's harder for attackers to exploit and safer for everyone to navigate. Your efforts are part of a larger movement, and every step you take contributes to building a more secure, trustworthy digital world.

Closing Words of Encouragement

Cybersecurity can feel overwhelming at times. The threats are real, the technology can be complex, and the stakes are high. But remember, you don't have to do everything all at once. Start small. Take one step today—whether it's setting up MFA, backing up your data, or scheduling a team training session.

Each action you take is a victory, a step toward stronger defenses and greater peace of mind. And those small steps add up to big change.

You have the tools, the knowledge, and the determination to protect your business. Embrace the challenge, take action, and know that every effort you make today will lead to a more secure tomorrow. Your business deserves it, and the digital world needs it. You've got this.

Additional Resources for Continued Learning

Building and maintaining a strong cybersecurity foundation requires ongoing learning and access to reliable tools and support. Below is a detailed list of books, blogs, organizations, tools, and professional contacts to help you stay informed and well-equipped.

Books, Blogs, and Newsletters
Books

1. *Cybersecurity for Dummies* by Joseph Steinberg – A beginner-friendly guide to understanding cybersecurity concepts and protecting your business.

2. *The Art of Deception* by Kevin Mitnick – Explores social engineering tactics and how to defend against them.

3. *Hacking Exposed: Network Security Secrets and Solutions* by

Stuart McClure et al. – A comprehensive guide to understanding various cyber threats and defensive strategies.

Blogs

1. **Krebs on Security** – Written by cybersecurity journalist Brian Krebs, this blog provides insights into emerging threats and trends.

2. **Dark Reading** – A trusted source for news and analysis on cybersecurity risks, technologies, and strategies.

3. **The Hacker News** – Covers global cybersecurity incidents, solutions, and best practices.

4. **BizTek Connection CyberBLOG** – https://biztekconn ection.com/blog/

Newsletters

1. **CISA Weekly Bulletin** – Updates from the Cybersecurity and Infrastructure Security Agency (CISA) on the latest alerts, threats, and resources.

2. **The CyberWire** – A daily newsletter offering concise updates on cybersecurity news and trends.

3. **SANS NewsBites** – Curated by the SANS Institute, this newsletter highlights important cybersecurity developments.

Cybersecurity Organizations and Online Communities

1. **Cybersecurity and Infrastructure Security Agency (CISA)** – Offers free resources, tools, and guidance tailored for small businesses.Website: www.cisa.gov

2. **National Institute of Standards and Technology (NIST)** – Provides cybersecurity frameworks and guidelines for organizations of all sizes.Website: www.nist.gov

3. **CyberAware** – A nonprofit initiative focused on raising awareness about online security threats and solutions.Web site: www.cyberaware.org

4. **Reddit Cybersecurity Community** – An active online forum where professionals and enthusiasts discuss topics, share advice, and recommend resources.Subreddit: r/cybersecurity

5. **LinkedIn Groups** – Groups like "Small Business Cybersecurity Forum" and "Cybersecurity Professionals" are great for networking and learning.

With these resources, you have access to the knowledge, tools, and support needed to strengthen your cybersecurity defenses. By staying informed and connected, you'll be well-prepared to protect your business in today's ever-changing digital landscape.

Conclusion: Building a Secure Future Together

As we come to the end of this journey, take a moment to reflect on the progress you've made. By committing to learning about cybersecurity, you've already taken a significant step toward protecting your business, your team, and your customers. This commitment demonstrates not only your dedication to your business's success but also your role in fostering a safer digital environment for everyone.

Reflecting on the Journey to Better Cybersecurity

Cybersecurity might have seemed like an overwhelming topic at the start, filled with jargon, complex tools, and endless threats. But look at where you are now—equipped with practical knowledge, actionable strategies, and a clear plan to safeguard your business.

This journey isn't just about tools and technology; it's about the mindset you've cultivated. You've taken the time to understand the risks, learn how to address them, and build a foundation of security that will grow with your business. Your efforts matter, and the steps you take today will ripple outward, protecting not only your business but also the people and partnerships that rely on it.

The Importance of a Growth Mindset in Cybersecurity

Cybersecurity is not a destination—it's a journey. Threats will continue to evolve, and technology will keep advancing. But that's where your growth mindset comes into play. By remaining adaptable and open to learning, you'll be ready to face new challenges with confidence and resilience.

Patience is key. Implementing changes, training your team, and integrating new tools takes time. Progress may be incremental, but every step you take strengthens your defenses. Remember, the goal isn't perfection; it's continuous improvement. With each update, training session, or tool you implement, you're building a business that's more secure, adaptable, and prepared for the future.

A Message of Empowerment

It's easy for small businesses to feel overshadowed in the cybersecurity world, where headlines often highlight the resources of major corporations. But never underestimate the power of your actions. No business is too small to make an impact.

Every password you strengthen, every phishing attempt you block, every employee you train—these are victories. They are proof that you can protect your business, no matter the size of your team or budget. You are part of a larger movement, one where businesses of all sizes contribute to a safer, more trustworthy digital world.

The power to make a difference is in your hands. By acting decisively and consistently, you're not just securing your business—you're leading by example.

Looking to the Future with Confidence

The digital world will always have its challenges, but you now have the tools, knowledge, and confidence to meet them head-on. You've built a foundation for a more secure future, one that's adaptable, resilient, and focused on continuous growth.

As you look forward, remember that cybersecurity is not just about defense—it's about opportunity. With strong security practices, you can innovate without fear, build trust with your customers, and focus on what matters most: growing your business.

The future is bright, and you're ready for it. Together, let's build a secure, thriving digital world, one step at a time. You've got this.

AFTERWORD

As you close this book, my hope is that you feel empowered to take meaningful steps toward protecting your business from cyber threats. Cybersecurity is a journey—one that evolves alongside technology, the threat landscape, and your business's unique challenges. The good news is, you don't have to navigate this path alone.

Let me introduce you to **BizTek Connection, Inc.** (www.biztek connection.com)—a trusted leader in cybersecurity and IT technologies. For over 24 years, BizTek Connection has been helping businesses of all sizes build strong, reliable defenses against cyber threats while optimizing their IT systems for maximum performance and security.

What sets BizTek Connection apart is our unmatched expertise. Our leadership brings over **71 years of experience in the IT field**, providing a deep understanding of how technology can empower businesses. Combined with a team boasting more than **106 years of collective expertise**, we offer a level of knowledge and insight that ensures your cybersecurity and IT solutions are tailored, effective, and cutting-edge.

Why BizTek Connection?

Cybersecurity isn't just about tools or technology—it's about expertise, strategy, and staying ahead of the game. At BizTek Connection, we take a proactive and personalized approach to IT management and cybersecurity.

Comprehensive Cybersecurity Solutions: From advanced threat detection to real-time monitoring and rapid incident response, we deliver end-to-end protection for your business.

AI-Powered Security: Harnessing the latest AI technologies, we implement intelligent, adaptive defenses designed to counter emerging threats.

IT Optimization and Support: Beyond cybersecurity, we focus on optimizing IT infrastructures to ensure your systems operate efficiently, securely, and in alignment with your business goals.

A Partner for Your Business's Unique Needs

At BizTek Connection, we pride ourselves on crafting solutions that are as unique as the businesses we serve. Whether you're a small business just beginning to address cybersecurity or an established company seeking advanced solutions, we work alongside you to develop strategies that meet your specific needs and budget.

Our ultimate goal is to give you peace of mind, so you can focus on what you do best—running and growing your business. With BizTek Connection, you're not just investing in technology; you're investing in a partnership built on decades of experience, a commitment to excellence, and a focus on your success.

The Journey Ahead

Cybersecurity is no longer optional—it's essential for thriving in today's digital business environment. At BizTek Connection, we believe in empowering businesses to embrace the future with confidence. Whether you need help implementing AI-powered tools, addressing vulnerabilities, or managing your IT systems, we're here to guide you every step of the way.

Visit to learn more about how our expertise and personalized solutions can support your business. Together, we can build a secure, resilient, and thriving future.

Thank you for joining me on this journey through cybersecurity. Here's to a safer, smarter tomorrow.

Acknowledgement

I'd like to take a moment to acknowledge something that might feel a bit self-serving—I am the founder of BizTek Connection, Inc. But as I wrote this book, sharing insights and strategies to help small businesses strengthen their cybersecurity, I realized it would be a missed opportunity not to introduce you to the services we offer.

BizTek Connection isn't just a business to me; it's a reflection of my passion for helping companies like yours succeed in an increasingly digital world. With decades of experience in IT and cybersecurity, my team and I have seen firsthand how the right technology can transform businesses, providing security, efficiency, and peace of mind.

I hope this introduction doesn't come across as overly promotional but rather as an extension of my desire to support you beyond the pages of this book. If anything you've read here feels overwhelming or out of reach, I want you to know that there's a team ready to help. BizTek Connection exists to make cybersecurity and IT management accessible, effective, and tailored to your unique needs.

Thank you for trusting me enough to read this book. Whether you choose to partner with BizTek Connection or take the tools and strategies here to build your own path, I'm honored to have been part of your journey toward a more secure future.

Consult the Experts

This book has been written as a practical guide to help small businesses understand the risks and strategies associated with cybersecurity. It is meant to provide foundational knowledge, actionable insights, and a framework for building stronger defenses. However, it's important to recognize that cybersecurity is not a one-size-fits-all solution. Every business has unique needs, vulnerabilities, and operational nuances that cannot be fully addressed through a general guide.

While the tools, tips, and strategies shared here can help you take meaningful steps toward protecting your business, they are not a substitute for professional analysis and expertise. Cybersecurity threats evolve rapidly, and only a qualified IT professional can provide tailored recommendations and implement comprehensive defenses suited to your specific situation.

If you have concerns about your current security posture or need assistance implementing advanced solutions, we strongly encourage you to consult a reputable IT company or managed service provider. Their expertise can ensure that your defenses are configured correctly, vulnerabilities are addressed proactively, and your business remains resilient against the ever-changing landscape of cyber threats.

Your business, your customers, and your peace of mind are worth the investment. Seek professional guidance to complement the knowledge you've gained here, and take your cybersecurity strategy to the next level.

.

ABOUT THE AUTHOR

ROGER BEST

MEET ROGER BEST

Roger Best is a seasoned entrepreneur who's navigated the wild ride of building multiple successful businesses. With years of battle scars and triumphs, he knows firsthand the relentless grind of entrepreneurship—those long hours, high-stakes decisions, and the relentless chase for success that can sometimes steal the joy out of life.

But Roger isn't just a survivor; he's a thriver. After his own transformation from a business owner burning the candle at both ends to a man who's mastered the balance of work and play, he's on a mission to help others break free from the chains of the endless hustle. In this book, he lays out practical, no-nonsense strategies to reclaim your time, achieve financial freedom, and craft a life packed with purpose, adventure, and a healthy dose of leisure.

When Roger's not in the trenches with one of his ventures, writing, or mentoring fellow entrepreneurs, you'll find him living life on his terms—spending quality time with family and friends, exploring new terrain, or just kicking back and enjoying the satisfaction that comes from running a successful business that doesn't dominate every moment.

Roger's been married to his soulmate for just over 45 years, a proud dad to two grown kids, and a devoted grandfather to two little princesses. In 2021, he and his wife decided to turn their island dream into reality, moving to Puerto Rico, where they now run their businesses while soaking up the hammock life.

Roger's mission is straightforward: to inspire and empower entrepreneurs to build thriving businesses without sacrificing their happiness, health, or freedom.

ALSO BY ROGER BEST

Explore a collection of inspiring and practical reads, authored by Roger Best, and designed to empower entrepreneurs and business owners. From reclaiming your time to building a business that truly serves you, these books offer actionable insights to help you create a successful and meaningful life and business.

- **From Hamster Wheel to Hammock: A Guide to Taking Your Day Back** *(Available on Amazon)* Escape the relentless grind and rediscover the freedom you dreamed of when you started your business. This practical and empowering guide offers actionable strategies to help you break free from the hamster wheel, reclaim your time, and build a balanced, fulfilling life. Whether you're overwhelmed by long hours or stuck in the cycle of constant hustle, this book will show you the path to a life of purpose and peace.

  - https://www.amazon.com/dp/B0DGXVW8D4

- **Personal Growth for Entrepreneurs: Your Time, Your Way** *(Available on Amazon)* Take a deeper dive into aligning your business and life with your values and passions. This book focuses on developing habits, mindsets, and systems that not only support your professional goals but also nurture your personal growth. Learn how to create meaningful success that feels true to who you are while finding fulfillment in every step of the journey.

○ https://www.amazon.com/dp/B0DLWXPK4L

- **The Liberated Entrepreneur: Building a Business That Works for You** *(Available on Amazon, as well as many other websites)* Running a business doesn't have to mean sacrificing your life. In this transformative guide, you'll learn how to build a business that aligns with your values, supports your dreams, and works *for* you—not the other way around. Discover strategies for streamlining operations, empowering your team, and creating a sustainable work-life balance. This book is for entrepreneurs ready to escape burnout and reclaim the joy and freedom they set out to achieve.

- **Purpose-Driven Entrepreneurship: Reclaim Your Why for Lasting Success** *(Available on Amazon, as well as many other websites)* Discover the power of reconnecting with your purpose and unlock the clarity and passion needed to thrive. This book offers practical exercises, journaling prompts, and reflective tools to help you rediscover your core motivations, make aligned decisions, and build a life and business that truly matter. Perfect as a companion to *From Hamster Wheel to Hammock* and *Personal Growth for Entrepreneurs*, this book is essential for anyone seeking deeper connection, fulfillment, and long-term success.

- https://www.amazon.com/dp/B0DSQMCQB6

- **Protecting Your Business in the Digital Age: A Non-Technical Guide to Cybersecurity for Small Business Owners** *(Available on Amazon, as well as many other websites)* Cybersecurity doesn't have to be overwhelming or overly technical. In this straightforward guide, you'll learn how to protect your business from digital threats, safeguard your financial resources, and ensure your customer data stays secure. Designed specifically for small business owners, this book breaks down the essentials of cybersecurity into practical, easy-to-understand steps that anyone can implement. Whether you're a tech novice or simply want peace of mind, this guide will help you secure your business and its future.

  - https://www.amazon.com/dp/B0DTZ6G2GG

Each of these books offers unique insights into different aspects of entrepreneurial life. Whether you're seeking to break free from burnout, deepen your personal growth, reconnect with your purpose, or protect your business in a rapidly changing digital landscape, these reads are your guide to a more intentional, successful, and resilient journey. Explore them today and take the next step toward the life and business you've always envisioned!

www.ingramcontent.com/pod-product-compliance
Lightning Source LLC
Chambersburg PA
CBHW071540210326
41597CB00019B/3065